SPECIAL FORCES
GUERRILLA WARFARE MANUAL

SCOTT WIMBERLEY

PALADIN PRESS
BOULDER, COLORADO

Special Forces Guerrilla Warfare Manual
by Scott Wimberley

Copyright © 1997 by Scott Wimberley

ISBN 10: 0-87364-921-4
ISBN 13: 978-0-87364-921-6

Printed in the United States of America

Published by Paladin Press, a division of
Paladin Enterprises, Inc.,
Gunbarrel Tech Center
7077 Winchester Circle
Boulder, Colorado 80301 USA
+1.303.443.7250

Direct inquiries and/or orders to the above address.

PALADIN, PALADIN PRESS, and the "horse head" design
are trademarks belonging to Paladin Enterprises and
registered in United States Patent and Trademark Office.

Visit our Web site at www.paladin-press.com

CONTENTS

This book is respectfully dedicated to those who have or will put their lives on the line in pursuit of or in the defense of the rights and freedoms that each of us inherits at birth, but which is often robbed or denied by others.

PREFACE

I started writing this book to help some friends in a far-away country regain their freedom. My motivation to write it came about when I began to realize that freedom is a fleeting thing that, once lost, is difficult to regain.

I have seen places where people are killed by their own governments. Places where the rulers hold power by force and rob the country of its wealth. Places where there is no education for the children and where chemicals are used as a weapon when citizens take up arms to defend themselves from genocide.

The book is *not* intended as a training manual for terrorists, who are basically cowards. It is intended to help people who are living under oppression and for whom all other means of change have been exhausted or rendered impossible. It is for those people who have been denied their rights and freedoms that everyone, everywhere deserve.

Armed conflict must remain the last resort, because it means sacrifice, death, and destruction. As such, I ask any reader that contemplates such an undertaking to give careful thought prior to initiating such action.

This book is based on my recent research as well as my training during eight years in the U.S. Army Special Forces during the Vietnam War era. I do not claim to have a monopoly on this knowledge, and readers are encouraged to study other sources of information. I suggest the following reading:

- *Guerrilla Warfare and Special Forces Operations* (FM 31-21). This is a very good text published by the U.S. Army. It is available in many surplus stores and at gun shows. It deals with the basics of guerrilla warfare, but much of it is specific to the U.S. military.
- *Guerrilla Warfare* by Che Guevara (University of Nebraska Press, 1985). With an introduction and case studies by Loveman/Davies. This book gives practical examples of guerrilla warfare and is an excellent example of a "peasant-based" revolution.
- *Guerrilla* by Charles W. Thayer (Harper & Row, 1963). This scholarly effort is the most interesting book I have ever read on the subject.

To anyone reading this book who is involved in an armed struggle against oppression, this book was written for you. May God be with you.

CHAPTER 1

PRINCIPLES OF GUERRILLA WARFARE

Guerrilla warfare has been around as long as recorded history. Proponents of conventional warfare, with its fixed lines of battle and rigid theology, often look down on insurgent action as dirty banditry. Guerrilla warfare is different than conventional warfare in many ways, but it deserves attention and consideration as a serious military science.

Guerrillas never win wars, but their adversaries often lose them. A guerrilla army cannot stand toe-to-toe with a well-trained and equipped conventional army. Instead, it can weaken and demoralize that conventional army. This will often cause the government supporting the conventional army to collapse or make political changes that the insurgents demand.

Resistance, rebellion, or civil war begins in a nation where political, sociological, economic, or religious oppression has occurred. Such discontent is usually caused by a violation of individual rights or privileges, the oppression of one group by a dominant group or occupying force, or a threat to the life and freedom of the people. Resistance can also develop in a nation where the once welcomed liberators have failed to improve an intolerable social or economic situation. It can also be inspired deliberately by external sources against an assumed grievance.

Resistance may be either active or passive. Passive resistance can come in the form of smoldering resentment, which needs only leadership or a means of expression to mature to active resistance.

Some people join a resistance movement because of an innate desire to survive. Others may join because of ideological convictions. Regardless of initial motivation, all are bound together to fight a common enemy.

In this chapter, we will look at the concepts of guerrilla war. Bear in mind that the subject can be covered only partially here, since just as conventional warfare requires volumes to discuss effectively, so does guerrilla warfare.

UNITY OF PURPOSE

Often in an insurgent movement, various factions have differing opinions as to what the final outcome of the effort should be. To be effective, however, an insurgency must first and foremost have unity of purpose. It must have a goal that is worthy enough to unify and rally the squabbling factions.

There should be a written document that states the reasons for the insurgency and its common goals. It should also outline basic goals of the resistance once successful. Above all, it must address the reasons why the people should be willing to take part in the resistance.

This document needs to be developed jointly by the various leaders of the resistance. Unity of purpose *must* be demonstrated by the leadership. Political infighting among resistance leaders is divisive and demoralizing to the movement. Resistance movements succeed or fail on the caliber of their leaders.

In addition to motivation and purpose, a population must feel that there is a chance for success, or no effective resistance movement can be developed. Active participation in any resistance movement is influenced by its chance for success.

CHARACTERISTICS OF GUERRILLA WARFARE

Guerrilla warfare is characterized by offensive action. Guerrillas relay on mobility, elusiveness, and surprise. Other characteristics include civilian support, outside sponsorship, political aspects, and tactics.

Civilian Support

Mao Tse-tung said that the population is to the guerrilla as the water is to the fish. The success and survival of a guerrilla force depends on continuous moral and material support from the civilian population.

The local community is usually under intense pressure from antiguerrilla forces. Punitive measures such as reprisals, deportation, restriction of movement, and seizure of goods and property are conducted against supporters of guerrilla activity, making support dangerous and difficult. If the local populace has a strong will to resist, however, such enemy reprisals cause an increase in underground activities.

The civilian community may assist the guerrillas by furnishing supplies, recruits, and information; giving early warning of antiguerrilla operations; supporting evasion and escape; and other activities.

After the guerrilla force has established itself and is sufficiently strong, it may need to exert force upon certain elements of the civilian population to command their support, i.e., coerce indifferent or unresponsive portions of the population into supporting the movement.

Outside Sponsorship

Guerrilla operations are more effective when they are

3

sponsored by outside sources. A sponsoring power decides to support guerrilla forces when it feels that the guerrillas can make a significant contribution toward its own national objectives. This support can be political, psychological, logistical, and/or tactical.

Political Aspects

Guerrilla warfare has been described as being more political than military in nature. It is military in a tactical sense, but it is also political since a guerrilla movement stems from a local power struggle. It often has its roots in the oppressive policies of a central government.

Offensive Guerrilla Tactics

By recognizing its own limitations, the guerrilla force can hope for survival and eventual success. Initially, the force is usually inferior to the enemy in firepower, manpower, communications, logistics, and organization. It is equal, and often superior, to the enemy in the collection of intelligence and the use of cover, deception, and time.

Because of logistical and manpower limitations, guerrillas must never directly confront a large, organized, superior military force in a pitched battle. Instead, guerrillas must initially coordinate their attacks against such targets as isolated enemy outposts, small police units, tax collectors, rail systems, roads and bridges, and other weakly defended installations. This has the combined effect of building the confidence of the guerrillas and forcing the enemy to guard installations rather than conduct offensive operations. It also demonstrates to the population that the resistance is there and is a force to be reckoned with.

The basis of successful guerrilla combat is offensive action combined with surprise. During periods of low visibility, the guerrilla unit attacks, tries to gain a momentary advantage of firepower, executes its mission, and leaves the scene of

4

action as rapidly as possible. Normally, the unit does not operate in one area but remains highly mobile and varies its operations so that no pattern is evident. If possible, it strikes two or three targets simultaneously to divide the enemy's attention and fragment its reinforcement effort. If pursued by enemy reinforcements, the unit tries to have countermeasures planned in advance such as booby-trapped withdrawal routes, indirect fire support during withdrawal (e.g., snipers), ambushes, dispersion, escape across a border into a sanctuary, prepositioned supplies, and, if surrounded, a preplanned breakout operation.

NOTE: If a guerrilla unit is being pursued by an enemy unit, it must avoid obvious danger areas such as roads, trails, clearings, and especially water because the enemy will try to keep these under surveillance.

Defensive Guerrilla Tactics

The guerrilla force develops a three-ring security system. The first ring is provided by the underground, which provides intelligence on the enemy's long-term plans and objectives. The second ring is provided by the auxiliary force and sympathetic locals, who provide timely intelligence on enemy troop movements, new units in an area, new commanders, new equipment, weather suitable for operations, extensions of enemy outposts, increased patrolling and aerial reconnaissance, and increased intelligence activities against the resistance. (Underground and auxiliary forces will be discussed in detail later in this chapter.)

The third ring of security is provided by the guerrilla fighters themselves. This includes such things as actively patrolling carefully selected bases, limiting access to bases, and using observation and listening posts, early warning devices, multiple concealed withdrawal routes, communications security, preplanned withdrawal, and temporary defense operations.

5

Upon receiving intelligence that the enemy is planning counterguerrilla operations, the guerrilla force commander should increase his own intelligence effort, determine the disposition and preparedness of his units, and review plans to meet the anticipated enemy action. The commander must not be too quick to overreact to unprocessed information that could have been planted by enemy sources, since this could interrupt operations, cause unnecessary movement, and broadcast planned defensive operations. Instead, preplanned defensive operations should be rehearsed during periods when enemy activity is not anticipated.

In the event that the commander receives positive confirmation that the enemy intends to conduct a major counterguerrilla operation in his area, he may choose to evacuate his bases without delay.

NOTE: If the enemy is concentrating its resources into one area, guerrillas in other areas take the opportunity to take savage action against lightly defended or inferior enemy targets and lines of communications. This makes the enemy less likely to concentrate its forces for extended periods.

If the guerrilla force is threatened by a superior enemy, it withdraws. If need be, it breaks into smaller units and disperses or attempts a breakout operation. The unit should have a preplanned regrouping plan in case it has to disperse.

Here is an example of a preplanned defensive operation. Assume the guerrilla base is in very rough, heavily vegetated terrain. The guerrillas get early warning of an enemy offensive in the area. The enemy moves in artillery, armor, and a large number of troops.

The guerrillas move out of the area after caching supplies and leaving behind recon teams. After the enemy moves its elements in, other guerrilla units destroy bridges, mine roads, set ambushes, deploy snipers, attack with mortars, and use diversions, hit and run tactics, and booby traps. The enemy is forced into defensive positions but cannot get a fix on the guerrillas. Every time the enemy sends out patrols, it suffers casualties. The enemy tries to resupply/reinforce, but his supply lines are harassed constantly.

The enemy has to choose whether to withdraw or move reinforcements and supplies in by diverting large numbers of forces. If he does withdraw, he is harassed by the guerrillas as he leaves. The enemy has trouble removing his heavy weapons and personnel because of snipers and indirect fire, blown bridges, land mines, ambushes, and actions of auxiliary forces that blend in with the local population.

Breaking Out of Encirclement

An encirclement maneuver is the greatest danger to guerrilla forces. Once the enemy has succeeded in surrounding the guerrillas, he may take one of several courses of action.

The simplest action would be for the enemy commander to have his troops close in from all sides until the guerrillas are trapped in a small area, which is then assaulted. Differences in terrain and rates of movement, however, make it almost impossible for troops to advance without the possibility of gaps being formed in the ranks. In other cases, the enemy may decide to break down the original circle into smaller pockets and clear them one at a time. In this event, the guerrillas may break out as the enemy attempts to maneuver into new positions.

Perhaps the most difficult situation for the guerrillas to counter is an assault after encirclement has been accomplished. In this situation, enemy forces on one side of the encircled area either dig in or use natural obstacles to block all possible routes, while the forces on the opposite side advance, driving the guerrillas against the fixed positions. As the advance continues, enemy forces that were on the remaining two sides are formed into mobile reserves to deal with any breakouts.

A guerrilla commander must be constantly on alert for indications of an encirclement. If the commander receives indications that an encircling movement is in progress, such as the appearance of enemy units from two or three directions, he immediately maneuvers his forces to escape while the enemy lines are still thin and spread out and coordination between advancing units is not well established. By doing so, the guerrilla force either escapes the encirclement or places itself in a more favorable position to meet it.

If for some reason escape is not accomplished initially, movement to a ridge line is recommended. The ridge line affords observation and commanding ground and allows movement in several directions. The guerrillas wait on this high ground until periods of low visibility or another favorable opportunity for a breakout attempt occurs.

NOTE: Even though a breakout maneuver is planned, preparations are also made for a defense against an assault, as the initial breakout may fail, or the enemy may attack before the maneuver is executed.

In a breakout maneuver, two strong combat detachments precede the main body. The main body is covered by flank and rear guards. If gaps exist between enemy units, the combat detachments seize and hold the flanks of the escape route. When there are no gaps in the enemy lines, these detachments attack to create and hold an escape channel. The breakthrough is timed to occur during periods of poor visibility, free from enemy observation and accurate fire. During the attempt, any guerrilla units not caught in the circle make attacks against the enemy's rear to lure forces away from the main breakout attempt and help to create gaps.

After a successful breakthrough, the guerrilla force should increase the tempo of its operations, raising morale and making the enemy cautious about leaving his bases to attack the guerrilla areas.

If the breakout attempt fails, the guerrilla commander

divides his force into small groups and instructs them to escape through enemy lines at night. This action should be taken only as a last resort, as it means the force will be scattered and inoperative for a period of time, and the unit's morale may be adversely affected. Reassembly instructions should be issued before the groups disperse. These actions should be rehearsed along with other tactical training.

Another situation that should be considered is the possibility that the enemy force will be able to seal off the escape routes. In the event that this happens, the enemy may choose to wait for the guerrillas to run out of food and water, or they may assault immediately. If this situation appears probable, primary and alternate defensive positions should be prepared. Snipers and booby traps can be used to slow the enemy and inflict casualties. Recon patrols can help detect any changes in the enemy strategy. If the enemy assaults, the guerrillas fight from their primary defensive positions until a signal is given to move to the secondary positions. The method of withdrawal is preplanned to keep effective fires on the enemy while a portion of the guerrillas reposition. Terrain is used to channel the enemy into areas covered by the most effective weapons.

Much of the success of a defense depends on the capabilities of the enemy and support for the surrounded guerrilla force by friendly units outside the encirclement.

PHASES OF GUERRILLA WARFARE

The U.S. Army Special Forces defines three phases in guerrilla warfare. In the first phase, preparation, organization, and initial operations are conducted. In the second phase, the guerrilla movement is mobilized into a conventional force. The third phase is demobilization, when the guerrillas are disarmed and prepared to be loyal members of the population.

I think defining three phases oversimplifies a complex issue, but no matter how many ways you slice a pie, it is still a pie. I will discuss more phases that, in reality, are subsets of the three.

Preparation, Planning, and Organization Phase

During this phase, the insurgency forces and general population must be prepared for resistance. Preparing the population is primarily psychological. Sometimes the enemy does this for the insurgency due to their policies and actions. Sometimes propaganda and other measures are necessary.

Because of the strength of the enemy, security is of prime concern during this phase. The resistance should try to remain secret at this time so it can infiltrate governmental agencies, organize and establish intelligence networks, align and train various factions, and choose initial targets.

If a foreign power is to provide any type of aid to the resistance, or if the resistance is trying to obtain such assistance, this planning/organization phase is critical. The potential sponsor will be concerned about the potential for political embarrassment, compromise, and the chance of success. If the sponsor does not recognize such things as good planning, training, and organization, it will likely not give assistance.

If there is a potential sanctuary in another country, oftentimes there will be members in the government of that country who do not support the resistance nor agree with its cause. Hopefully they will be in the minority, but if that is not the case, these people must be dealt with or somehow be made ineffective so they are not a threat. Identifying these people during the initial stage is important so that plans are made to accommodate or otherwise deal with them. Sometimes political struggles in the potential sanctuary can be exploited. Those friendly to the resistance or the undecided need to be cultivated.

Initial Operations Phase

Combat operations are begun during this phase. The

types of targets chosen are those that almost guarantee success. Examples are unmanned targets with no enemy forces near, lone policemen, tax collectors, and small, isolated enemy outposts. Tactics include sniper operations, assassinations, and the use of easily placed explosives.

During this phase, protests against the government can be organized. Any action the government takes against protesters is used against it as propaganda. Governments often react with harsher policies and restrictions, which can be exploited further.

The guerrillas enhance ties with the local population by giving what they can spare in the way of food, medicine, labor, etc.

The desired effect of these operations is to give the guerrilla units experience in planning and conducting operations, instill confidence in the guerrillas, and show the population that the resistance can conduct successful actions against the enemy, thus providing encouragement to those wanting change. This also helps put teeth in psychological operations and helps in recruiting.

Expansion Phase

During this phase, small units of guerrillas start to confront armed enemy forces directly, attack defended positions, cut lines of communications, and expand logistics efforts, intelligence gathering, and communications capabilities.

Guerrilla units execute aggressive offensive operations such as ambushes, raids, and roadblocks. They increase their dominance over their areas of influence and make it increasingly more dangerous for the enemy to venture into those areas. They also establish and expand political and administrative control in their areas of influence.

Mobilization Phase

After the resistance is strong enough and has acquired a solid infrastructure of logistics and command, it transforms

11

its forces into a conventional army and starts conducting a conventional war by engaging large enemy units in ground combat. It attempts to gain the initiative in combat by establishing and maintaining contact with the enemy.

The time chosen to do this is critical. If it is done too soon, it exposes the resistance to enemy forces it cannot defeat. The implications of this can be militarily, psychologically, and politically devastating for the resistance. This mistake was made by the North Vietnamese in 1968 during the Tet Offensive. It almost cost them the war.

Political Consolidation Phase

This phase occurs at the end of a successful revolution, when the central government has lost its power. Long before the downfall of the government, however, the resistance leaders must have a plan to instill order that will consolidate power, determine which leaders will head the various sections of the new government, establish tentative rules of governance (including some type of court), and declare martial law if necessary. It is also beneficial to get some sort of international recognition by a major country.

If there has not been adequate preparation for restoring order once the old government has fallen, it could result in factional fighting for power, segments of the old regime reorganizing and resisting the new order, or a neighboring power moving in to fill the void.

Demobilization Phase

This is the time when the rifles are taken away from the guerrillas and they are given plows. If this is not done, there is a good chance that factions will fragment from the new government and start hostilities and/or the economy will become hampered, causing strife.

NOTE: In all types of war, discipline is very important. In guerrilla warfare, however, it is key. Every guer-

rilla leader in history has stressed the need for a highly organized, thoroughly disciplined core of dedicated fighters as a prerequisite for a successful guerrilla campaign. Romantic idealists who take to the jungle in order to fight despotism rarely are inclined to administer the kind of harsh justice required to deter the careless from inadvertent lapses of security or to steel the faint-hearted from revealing vital secrets under torture. Only after they have suffered costly casualties through the slips or weakness of their own colleagues do they learn the necessity for extralegal methods to enforce discipline among themselves and their civilian supporters. In fact, it is questionable whether a movement based on democratic organization and dedicated to legality can ever be ruthless enough to impose the brutal discipline a modern guerrilla force demands.

RESISTANCE ORGANIZATION

The organization of the resistance is key to the success or failure of the effort. Proper grouping of personnel, assignment of tasks and responsibilities, establishment of a chain of command, and organizing of security are paramount.

There are three organizational elements in the area of operations: the auxiliary, the underground, and the guerrilla force.

Auxiliary

Active support from some of the civilian population and passive support from most of the remainder is essential to extended guerrilla operations. Auxiliary forces provide for and organize civilian support of the resistance movement.

Auxiliary forces are composed of people who are not members of other resistance elements but who knowingly and willingly support the cause. It includes the occasional supporter as well as the hard-core leadership. Individuals or

13

groups who furnish support either unwittingly or against their will are not considered auxiliaries. Auxiliaries may be organized in groups or operate as individuals.

Auxiliary forces are characterized by their location, organization, and method of operation.

Location

Auxiliary units are composed of civilians normally living in smaller towns, villages, and rural areas. Unlike guerrilla units, the auxiliaries are not expected to move place to place to conduct operations. Auxiliary forces are local and static, which is highly desirable because it provides support for the mobile guerrilla forces throughout most of the operational area.

Organization

Auxiliary forces are normally organized to coincide or parallel the existing administrative divisions of the country—regional, county, or local (communities and villages). This method of organization ensures that each community and the surrounding countryside falls under the responsibility of an auxiliary unit. This organization can vary from country to country depending upon the existing political structure. Organization of auxiliary units can commence at any level or at several levels simultaneously and is either centralized or decentralized.

Basic organization at each level is handled by the command committee. This committee controls and coordinates auxiliary activities within its area of responsibility. In this respect, it resembles the command staff of a military unit. Members of the command committee are assigned specific duties such as supply, recruiting, transportation, communications, security, intelligence, and operations. At the lowest level, one individual may perform two or three of these duties.

The command committee may organize civilian sympa-

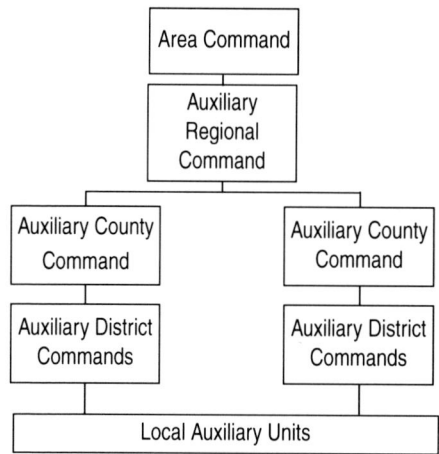

Figure 1. Centralized auxiliary organization.

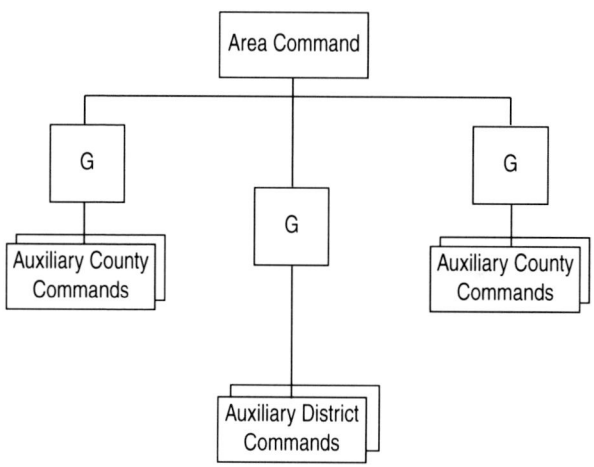

Figure 2. Decentralized auxiliary organization.

15

thizers into subordinate elements or employ them individual-ly. When possible, these subordinate elements are organized into a compartmented structure. But again, because of a shortage of loyal personnel, it is often necessary for each sub-ordinate auxiliary element to perform several functions.

The home guard is the paramilitary arm of the auxiliary. Home guards are controlled by the command committees. All auxiliary elements do not necessarily organize home guards. Home guards perform tactical missions, guard caches, and train recruits. Their degree of organization and training depends upon the extent of effective enemy control in the area.

Method of Operation

Auxiliary units derive their protection in two ways: orga-nizing a compartmented structure and operating under cover. While enemy counterguerrilla activities often force the guer-rilla fighters to move away from given areas temporarily, the auxiliaries survive by remaining in place and conducting their

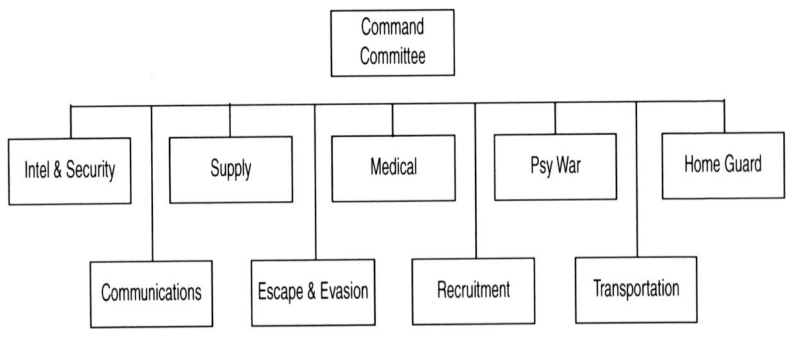

Figure 3. Possible organization of an auxiliary unit.

activities covertly. Individual auxiliary members carry on with their normal, day-to-day routines while secretly carrying out the many facets of resistance action.

Auxiliary units frequently call upon the passive or neutral elements of the population to provide active support to the common cause. Usually this is done on a one-time basis because of the security risks involved in repeated use of such people. The ability of auxiliary forces to manipulate large segments of the neutral population is enhanced by the demonstrated success of guerrilla forces.

Some of the support missions of the auxiliaries are coordinated directly with guerrilla units, while others are controlled by an auxiliary unit's higher headquarters. Normally, auxiliary units are assigned direct support missions for guerrilla units in their areas. Support missions include:

Security and warning. Auxiliary units provide a physical security and warning system for guerrilla forces. They organize extensive systems of civilian sympathizers who keep enemy forces under surveillance and warn the guerrillas of enemy activity. Candidates are assigned to the security system because of their advantageous location, which permits them to monitor enemy movement toward guerrilla bases.

Intelligence. The auxiliary force provides direct intelligence support to guerrilla units operating within their area of responsibility. They also collect information to support their own operations and those of the resistance in general.

Counterintelligence. The auxiliary supports the resistance counterintelligence effort by maintaining watch over transitory civilians, screening recruits for guerrilla units, and monitoring refugees and other noninhabitants in the area. Because of their intimate knowledge of local people, auxiliaries should be able to report attempts by enemy agents to infiltrate the area. They can also identify inhabitants whose loyalty to the resistance might be suspect.

Logistics. The auxiliary supports the guerrillas in all

17

aspects of logistic operations. They provide transportation and/or porters for movement of supplies and equipment. They often care for sick or wounded guerrillas, provide medical supplies, and arrange for doctors and other medical personnel. They establish and secure caches. They collect food, clothing, and other supplies for guerrilla units through a controlled system of levy, barter, or contribution. Sometimes auxiliaries provide essential services to the guerrillas such as repair of clothing, shoes, and certain items of equipment. Auxiliary units furnish personnel to assist at drop and landing zones, and they distribute supplies throughout the area. The extent of logistical support furnished by the auxiliary force depends upon the resources of the area, the degree of influence the auxiliary exerts on the population, and enemy strength and activities in the area.

Recruiting. The guerrillas depend upon the local population for recruits to replace losses and to expand their forces. Auxiliaries spot, screen, and recruit personnel for active guerrilla units. If recruits are provided through reliable auxiliary elements, the enemy's chances for placing agents in the guerrilla force are greatly reduced. In some cases, auxiliary units provide rudimentary training for guerrilla recruits.

Psychological Warfare. A very important mission in which auxiliary units assist is psychological warfare. The spreading of rumors, leaflets, and posters is timed with guerrilla tactical missions to deceive the enemy. Leaflets, for example, can mislead the enemy as to guerrilla intentions, capabilities, and location. The spreading of this propaganda usually involves little risk to the disseminator and is very difficult for the enemy to control.

Civilian Control. To control the population and give the enemy an impression of guerrilla power, the auxiliary units establish a rudimentary legal control system. This system can control black marketing and profiteering for the benefit of the guerrilla force. Collaborators may be terrorized or eliminated by the auxiliaries. In addition, auxiliary units can control large numbers of refugees in the area for the guerrilla force.

Evasion and Escape. Auxiliary units are ideally suited for the support of evasion and escape. Their contact with and control over segments of the population provide the area commander with a means of assisting evaders.

Other Missions. Auxiliary units may be called upon to provide a number of other missions to support guerrilla operations. Some of these are:

- Activity in conjunction with guerrilla military actions, such as cutting telephone lines between an enemy target and reinforcements prior to an attack.
- Operation of drop or landing zones.
- Operation of courier systems between widely dispersed guerrilla units.
- Furnishing guides to guerrilla units.
- Under some circumstances, conducting active guerrilla operations in their areas of responsibility on a part-time basis.

The Underground

Enemy security measures and/or apathy of some of the population often deny selected portions of an operational area to guerrilla forces or its auxiliaries. Since these areas are usually essential to the support of enemy operations, the resistance must attempt to extend its influence into them. The element that is used to conduct operations in these areas is the underground. The underground, then, is that element of the resistance established to reach targets not vulnerable to other elements. The underground is used to achieve objectives that would otherwise be unattainable.

In many ways, the underground closely resembles the auxiliary force. It conducts operations in a similar manner

19

and performs many of the same functions. The major differences are twofold: the underground is tailored to conduct operations in areas that are normally denied to the auxiliary force or guerrillas, and it does not relay as much on influencing the civilian population for its success.

Command Group

The command group provides a means to control and coordinate all resistance activities in a guerrilla warfare operational area. The group is normally located with the guerrilla force. In some cases, the command group may be located with the auxiliaries or underground.

SECURITY

Coincident with establishing a command organization in the guerrilla warfare operational area is the organization of adequate security. Security of all elements of the guerrilla movement is based upon preventing the enemy from either knowing of the existence of resistance forces altogether or, when the existence is known, preventing the enemy from locating these forces.

The guerrilla force normally utilizes a two-zone security system consisting of inner and outer security zones.

The inner security zone is the responsibility of the guerrilla units. In this zone, security depends on standard military techniques such as:

- Patrols.
- Outposts.
- A sentinel system.
- Warning devices.
- Cover and deception.

In the outer security zone, the auxiliary force and the

underground provide security for the guerrillas by furnishing timely information of enemy activity.

Since the underground and auxiliary forces achieve security by remaining undetected and through their basic cellular structure, this section deals only with the security measures applicable to the guerrilla force.

Principles of Security

Security is achieved by a combination of active and passive means, to include:

- Dispersion.
- Mobility of units and installations.
- Cover and deception.
- Records security.
- A physical warning system.
- March security.
- Counterintelligence activities.
- Support from the auxiliary and underground.
- Communication security.

Dispersion

Guerrilla forces avoid concentrating its troops in large camps or bivouacs. Even though the logistical situation may permit sizable troop concentrations, the force is generally organized into smaller units and dispersed. Dispersion facilitates concealment, mobility, and secrecy. Large forces may be concentrated to perform a specific operation, but upon completion of the operation, they disperse quickly.

The principle of dispersion is applied to both command and support installations. A large guerrilla headquarters, for example, is divided into several echelons and deployed over a large area.

In the event of well-conducted, large-scale enemy operations against the guerrilla force, the area commander may divide units into even smaller groups to achieve greater dispersion and facilitate escape from encirclement. Splitting units into such small groups is used only when all other means of evasive action are exhausted. Extreme dispersion reduces the effectiveness of the force for a considerable period of time. It also lowers the morale of the guerrillas and weakens the will of civilians to resist. To increase the probability of successful reassembly of dispersed units, preplanned primary and alternate assembly areas are established.

Mobility

Guerrilla forces and installations must maintain a high degree of mobility. Evacuation plans include elimination of all traces of guerrilla activity prior to abandonment of the area. Evacuation operations should be rehearsed.

Mobility for evacuation is achieved by preparing equipment to be moved in one-man loads, caching less mobile equipment, destroying or hiding material of intelligence value to the enemy, policing the area, and eliminating signs of the withdrawal route.

NOTE: Premature or unnecessary movement caused by the presence of the enemy may expose guerrillas to greater risks than remaining concealed. Such moves disrupt operations and tend to reduce security by exposing guerrillas to enemy agents, informants, and collaborators. The decision by the guerrilla commander to move is made after a careful evaluation of the situation.

Cover and Deception Operations

Deception operations deceive the enemy as to location, intent, or strength of the guerrilla force. They can be conducted in conjunction with other resistance operations in the area. Some examples of cover and deception operations are:

Phony Radio Transmission. Small teams make phony radio transmissions near an enemy installation to deceive the enemy into anticipating an attack and possibly diverting forces to reinforce the phony target, while other forces mass against another target or ambush the reinforcements.

Probes. Small teams probe enemy positions to evaluate enemy defenses and determine weak points. This is done over an extended period of time on an occasional basis to harass and occupy the enemy forces and deceive them as to when the preliminary actions of a real attack are taking place.

Phony Intelligence. Guerrillas plant phony intelligence material in a phony base camp so it can be found by the enemy. This is particularly effective in a closed society, where there is distrust among governmental groups or individuals.

Safeguarding Plans and Records

Information concerning guerrilla operations is disseminated on a need-to-know basis. Each person is given only that amount of information that is needed to accomplish his task. Special efforts are made to restrict information given to individuals who are exposed to capture. Minimum copies of documents are made.

Administrative records are kept to a minimum and cached so that only a required few know of their location. Essential records can be photographed to archive them; the film can be exposed to light if compromise is eminent.

Whenever possible, references to names and places is coded and the key to the code is given on a need-to-know basis.

Records of no further value are destroyed.

The guerrilla relies upon his memory to a greater extent than regular soldiers. Installations and operational details are not marked on maps and papers that leave the guerrilla base.

Security Measures

Strict security consciousness is impressed upon guerrilla

troops at all times. Commanders at all levels constantly strive to improve security measures, and all such measures are enforced. These include but are not limited to:

Camouflage Discipline. This includes an individual soldier's camouflage (e.g., noncontrasting clothing, breaking up a weapon's outline) and guerrilla base concealment.

Isolation of Units. Isolating units from each other helps prevent operational information from one unit being passed on to other units.

Courier Route Security. Proper selection and rigid supervision of courier routes between headquarters and units.

Keeping Camp Sites and Installations Clean. No dropping of any trash or discarding of materials that could give an indication of guerrilla presence or otherwise be of intelligence value.

Interbase Security. Movement control within and between guerrilla bases and installations.

Isolation from the Civilian Population. Any necessary contact with civilians is accomplished with auxiliaries. Troops are warned of the dangers of discussions with any civilians, including members of their families. Innocent talk can have disastrous consequences. Consideration should be made to isolate troops before they are allowed to visit family members.

Anti-Interrogation Methods. Thorough indoctrination of all resistance units to enemy interrogation methods and how to counter them.

Preplanned Actions in the Event of Capture. Any individual who does not return from an operation or whose body is not recovered must be considered captured. Any information that individual knew or possessed is considered compromised. Individuals are trained and rehearsed on what to do if separated from their unit. The guerrilla force moves to a predetermined but undisclosed alternate base, and the old base is kept under surveillance for a given amount of time to watch for missing personnel. Upon returning to the old base, the individual must give some sort of signal to authenticate the fact that he was not captured and forced to reveal the base's location. Once the surveillance personnel are sure the individual was not captured or followed, they make contact with him.

March Security

Security on the march is based upon accurate knowledge of the enemy's location and strength. The intelligence section of the resistance usually provides this vital information. The unit commander should also consider conducting reconnaissance to update/enhance his unit's knowledge of the enemy.

Once routes have been selected, units are briefed on primary and alternate routes, enemy activity, dispersal and reassembly areas along the way, and security measures to be used en route. Auxiliary units in the route area assist by providing security elements for the guerrillas as well as any additional information on the enemy they have gained.

While on the move, guerrilla forces employ march security techniques such as advance, rear, and flank guards. Preselected bivouacs are thoroughly screened by patrols prior to their occupation by guerrilla forces.

Counterintelligence

Security measures used by guerrillas to safeguard information, installations, communications, and forces are supplemented by an active counterintelligence program to neutralize the enemy's intelligence system and prevent the penetration of guerrilla forces by enemy agents.

Counterintelligence is a command responsibility under the supervision of the intelligence section. Selected personnel, specially trained in counterintelligence, carefully screen all members of the guerrilla organization as a protective measure against enemy infiltration. They also plan and supervise an active campaign of deception.

25

Counterintelligence personnel within the auxiliary force monitor the civilian population constantly to watch for the presence of enemy agents within their midst. Civilians upon whom the guerrillas depend heavily for support may compromise the guerrilla effort as easily as disloyal guerrillas.

False rumors and false information concerning guerrilla strength, location, operations, training, and equipment can be disseminated by counterintelligence personnel. Facts are distorted intentionally to minimize or exaggerate guerrilla capabilities at any given time.

Active measures are taken to determine enemy intentions and methods of operation and to identify enemy intelligence personnel or local inhabitants who may be acting as enemy agents. These measures include penetration of enemy intelligence and counterintelligence organizations by selected personnel and the manipulation of defectors and double agents.

Security Role of the Auxiliary and Underground

Both the auxiliary forces and the underground contribute to the security of the guerrilla force by operating in what is to the guerrillas the outer security zone. Incidental to their everyday operations, they uncover enemy activity or indications which, when evaluated, disclose potential danger to the guerrilla force. They establish specific systems designed to provide warning of approach of enemy units. They intimidate any collaborators and attempt to elicit information from enemy personnel, local officials, and police.

COMMUNICATIONS

Communications within an area furnishes the commander with a means of controlling his forces. Because the guerrilla force often operates in an area dominated by the enemy, its communication system will be slower than the enemy's.

More time must be allowed for transmitting orders than in a conventional warfare environment.

Whenever practical, guerrilla commanders use nonelectronic means to communicate. Unless an area is relatively secure, electronic means should be used only when absolutely necessary due to the possibility of the use of radio direction-finding equipment by the enemy.

Messengers

During the early phases of an insurgency, messengers are the primary means of communications. Security is enhanced by establishing a cellular structure for the messenger organization and through the use of codes and encryption.

Messengers should vary their routes so they do not set patterns that could be exploited by the enemy. They should use predetermined recognition signals and change them often. They should not know the content of any coded messages nor the identity of the recipient if possible.

Messengers are instructed not to talk with other messengers in their cell concerning the contents of messages, routes, or other operational details. Their knowledge of guerrilla bases, installations, and safe houses must be limited.

Radio

Radios should be used only by trained personnel. Persons trained in communications security operate radios using strict radio discipline. They use terrain masking, directional antennas, brevity codes, call signs, and physical security techniques while operating radios.

Radio transmissions should never be done at or near a friendly base unless the situation dictates it. If this is done, the commander must assume that the enemy will determine the location of the transmission.

The range of radios, which operate in the high-frequency range, is difficult to predict. Under ideal conditions, these transmissions can be intercepted over great distances by most government forces. The range of low-powered radios operating in the VHF band rarely exceeds line of sight, but the type of antenna used determines the transmission pattern and relative signal strength in a given direction.

Often a transmission site can be located across a border and used to relay messages or transmit "blind transmissions." Blind transmission broadcasts (BTB) are transmissions made to deliver information, orders, or operational details to guerrilla units. The recipients of BTBs just listen and never acknowledge receipt of messages. A BTB message may be sent twice in a row to ensure reception. The transmissions are prescheduled, and it is important that all BTBs are coded and authenticated. Dummy BTBs can be transmitted, or the messages may be repeated at different times of the day. BTBs are often targeted by the enemy using jamming equipment.

Packet radio is a fairly new capability available to the guerrilla. It currently requires a radio and a personal computer as well as the proper software. The advantage of packet radio is that a message can be created on a computer, then transmitted at high speed. This causes the transmission time to be much less than that of voice, which reduces the available time an enemy has to get a "fix" on the transmitter.

Packet radio was originally limited to transmission rates of 1200 baud, but transmission rates should be increasing dramatically in the near future due to the popularity of this medium. Manufacturers of commercially available radios should soon produce equipment that allows higher rates of data flow.

Encryption software is available for personal computers. One such program is PGP (which stands for "Pretty Good Privacy"). When encryption is coupled with the speed of packet radio, VHF line of sight, directional antennas, and terrain masking, radio security is increased dramatically.

28

LOGISTICS

There are three primary sources of logistical support for guerrillas. The first and usually the most important is that which can be obtained from the local area. This includes both the local population and the land. The local population often provides food and supplies to support the guerrillas, but this should never be taken for granted because the support of the population is extremely important to the insurgency. Whenever possible, the supplies are paid for in some way or an I.O.U. is given and not forgotten.

The second source of logistical support is an outside sponsor. The sponsor can often provide supplies and equipment that cannot be obtained locally or in quantities exceeding those available locally. Usually, the biggest problem with logistical support from external sources is getting the supplies to where they are needed. There is often a border to cross, and enemy forces will most likely patrol this border. Also, the uncertainties of weather and enemy action prevent timetable receipt of supplies from the sponsoring power.

The third source of logistical support is the enemy. When possible, guerrillas should capture such things as food, weapons, ammunition, medical supplies, and other supplies and equipment from the enemy. Operations can be conducted with the primary purpose of obtaining supplies, or they can be gathered after an operation has been conducted for other reasons. The auxiliary force is often tasked to transport and/or store captured supplies.

Regardless of how the bulk of supplies are obtained, the area command must have an element tasked with the responsibility of logistics and transportation. This group is responsible for procurement, planning, training, and oversight of logistics and transportation.

The area command can assign each guerrilla unit a portion of the operational area for logistical support. Usually the guerrilla units receive direct logistical support from the auxiliary units within their assigned portion of the operational area. In addition to support from local auxiliaries, the guerrilla unit depends on its own overt actions to satisfy logistical requirements.

One of the primary responsibilities of the auxiliary units is logistical support of guerrilla units. Since the auxiliaries themselves are largely self-sufficient because they live at home, they establish local systems designed to support guerrilla units.

The underground logistical role is largely one of self-maintenance for its own members.

Caching

Storage or caching of supplies and equipment plays an important role in the area command logistical plan. The area command must be prepared to operate for extended periods without external resupply. This necessitates stockpiling supplies for later use. Guerrilla units do not maintain excess stocks of supplies since large quantities of equipment limit mobility without increasing combat effectiveness.

Supplies in excess of current requirements are cached in a number of isolated locations to minimize the risk of discovery by the enemy. These caches are established and secured by both guerrilla and auxiliary units. Items are packaged carefully so that damage from weather and exposure is minimized.

Caches can be located anywhere that material can be hidden—caves, swamps, forests, cemeteries, and lakes. The cache should be readily accessible by the guerrilla. Dispersal of the caches throughout the operational area permits a high degree of flexibility for the guerrilla force. Only the commander and key personnel know the locations of the caches.

Transportation

Transportation for supplies is acquired primarily within

the area of operations. Movement by foot is usually the primary means available to the guerrilla in the initial stages of development. In special situations, this may be supplemented by locally procured motor vehicles, bicycles, or animals. The auxiliaries provide whatever local transportation is available, usually on a mission basis.

In some instances, guerrillas can permanently acquire transportation. For example, in a desert warfare situation, guerrillas may need more mobility than foot patrols allow. Acquiring vehicles for this purpose depends upon such things as enemy air observation capabilities and the guerrilla's capacity to refuel and maintain vehicles. Guerrillas operating in an urban environment could also use vehicles more readily than their rural counterparts.

MEDICAL SERVICE

Normally, medical services available to guerrillas is that which is available in the area of operations. It is sometimes augmented by medical supplies provided by the sponsor.

Due to the nature of guerrilla warfare, battle casualties are usually lower than that of conventional infantry units. The incidence of disease, however, is often higher than conventional units.

The medical system in the operational area features individual combat medics, organized medical units, and auxiliary medical facilities. The organization of the medical detachment consists of three sections: the aid station, which is charged with the immediate care and evacuation of casualties; the hospital, which performs defensive treatment of casualties and coordinates training and resupply; and the convalescent section, which cares for patients who require rest and a minimum of active medical attention before their return to duty.

The convalescent section is not located near the hospital since this would increase its size and create a security risk.

Instead, patients are placed in homes of local sympathizers or in isolated convalescent camps. During initial stages of development, aid stations may be set up in the same installation as the hospital.

Every effort is made to evacuate wounded personnel from the scene of action. The condition of wounded guerrillas may preclude their movement with the unit to the base. In this event, the wounded are hidden in a covered location and the local auxiliary force notified. The auxiliaries then care for and hide the wounded until they can be returned to their units.

The evacuation of the dead from the scene is extremely important for security reasons. The identification of the dead by the enemy can jeopardize the victim's family as well as his unit. The bodies of those killed are evacuated and cached until they can be recovered for proper burial, or they are disposed of by whatever means is consistent with the customs of the local population.

COMBAT IN THE GUERRILLA
WARFARE ENVIRONMENT

Guerrilla warfare is characterized by constant change. There are no fixed lines between the opposing forces. Maximum effective results are attained by the guerrilla force through offensive operations. Normally, the guerrilla force is primarily interested in the interdiction of lines of communications and the destruction of critical enemy installations. Except for those instances where the tactical advantage is clearly with the guerrilla force, no effort is made to close with and destroy the enemy. The enemy, on the other hand, must provide security for its critical installations and seek contact to destroy the guerrilla force.

Guerrilla forces are rarely concerned with seizing and holding terrain. They are concerned with controlling areas, however. Controlling an area offers greater security and support for the resistance, facilitates operations, and provides a location from which a political stance can be voiced.

Area control is classified in two basic ways: area superiority and area supremacy.

Area superiority involves temporary control of a specific area by the use of the principles of surprise, mass, and maneuver. Area superiority is maintained only for the period of time required to accomplish missions without prohibitive interference from the enemy.

Area supremacy involves complete area control when the enemy is incapable of effectively interfering with guerrilla operations. Area supremacy is seldom achieved by a guerrilla force until it progresses to the conventional warfare stage.

Guerrilla area superiority is more easily achieved in difficult terrain that restricts enemy movement and observation. These factors reduce the enemy's ability to find, fix, and maneuver against the guerrilla force. Difficult terrain also allows more time for the guerrillas to escape enemy assaults, thus avoiding fighting a defensive battle. Areas of sustained guerrilla superiority are best located away from critical enemy targets.

The enemy is usually free to establish superiority of any area it chooses as long as it is willing to commit sufficient forces to do so. However, because the guerrilla force is comparatively free to select the time and place of attack, successful operations against targets are conducted despite this enemy superiority.

Between the areas where guerrillas have strongholds and the enemy has installations is a twilight zone that is never under complete control by either. Because of the mobility of the guerrillas, their detailed knowledge of the terrain, and their ability to choose targets selectively across widespread areas, security of this twilight zone by the enemy is impossible.

While the enemy and the guerrillas compete for overt control throughout the twilight zone, guerrillas cannot hold any specific area against determined enemy attack. The enemy holds those areas which it occupies by force, and the guerrillas conduct operations in those areas where the enemy is weakest. The auxiliary is more effective in the twilight zone than it is in the enemy-dominated areas. Intelligence organizations report everything the enemy does in the twilight zone. Throughout the twilight zone, the enemy is made to feel it is in hostile territory—it may control a small portion by force of arms, but it can never relax its guard lest it be surprised by the guerrillas.

Guerrilla operations and intelligence activities in the twilight zone aids in the security of the areas of guerrilla superiority because it occupies enemy forces, often allows the resistance advanced knowledge of enemy movements, and places the guerrillas in areas where they can undertake offensive operations against targets left vulnerable by the enemy. When the enemy does attack, the guerrillas usually do not attempt to defend—rather they withdraw, create diversions, and attack the enemy's flanks and rear areas.

CHAPTER 2

SMALL UNIT TACTICS

The purpose of this chapter is to introduce the fundamentals of small unit tactics. Successful small unit operations are based on principles developed in many wars over many years. These principles form a basis for platoon and squad tactics, techniques, procedures, and drills. Also discussed are the elements of combat power and the skills required of leaders and soldiers at the small unit level.

General concepts of infantry tactics are covered as they apply to insurgent warfare, and specific concepts are discussed that are of special concern to the insurgent.

NOTE: Understanding the concepts introduced in this chapter is important for the individual soldier and small unit leader, but there is no substitute in combat for common sense, flexibility, ingenuity, and planning. Training is important to the soldier. Practice is just as important. The unit must rehearse potential actions and become a team. It is not enough to simply read and understand tactics. It must become second nature.

MISSION

The mission of the infantry is to close with the enemy by means of fire and maneuver to defeat or capture him, or repel his assault by fire, close combat, and counterattack.

NOTE: In insurgent warfare, the mission includes the above; however, the insurgent must close with the enemy *only* when the situation is to his advantage. Often, the central government has greater firepower that it can mass as well as air power. The advantages the insurgent has are the ability to remain mobile and hidden and strike when and where he chooses, then disperse before the enemy can mass its forces for attack. This requires very good intelligence about the enemy, a high degree of training, support from the local population, and making maximum use of terrain.

Despite any technological advantages the enemy may have over insurgent forces, only close combat between ground forces gains the decision in battle. The insurgent can gain the advantage in a tactical situation by:

- Attacking over approaches that are not possible for heavy forces.
- Using sudden violent attacks against smaller forces in isolated locations.
- Restricting enemy movement and resupply by cutting lines of communications.
- Attacking targets that restrict the enemy's ability to respond to attack and/or secure rear areas (e.g., fuel supplies, vehicles, aircraft, communications facilities, government facilities).
- Attacking targets that require the enemy to deploy large numbers of forces to secure them, such as government buildings, power plants, and bridges.
- Seizing or securing forested (jungle) and built-up areas.

The insurgent force usually does not attempt to hold ground if a concentrated attack is launched against it. It does however, often make it costly for the enemy to attack such areas and difficult for them to hold.

- Exploiting the advantage of surprise.
- Concentrating on planning. The small unit *must* concentrate on planning. Planning is the design of an operation, and without it, the units are just armed personnel that will not be effective.

NOTE: The successful actions of small units rely on the ability of leaders and soldiers to use the terrain to good advantage, to operate their weapons with accuracy and deadly effect, and to out think, out move, and out fight the enemy.

COMBAT POWER

The doctrine that guides small unit forces is based on the four elements of combat power: maneuver, firepower, protection, and leadership.

Maneuver

Maneuver is the movement of forces while supported by fire. It is done to achieve a position of advantage from which to destroy or threaten destruction of the enemy. Properly supported by fires, maneuver allows the infantry to close with the enemy and gain a decision in combat. Infantry forces maneuver to attack enemy flanks, rear areas, logistics points, and combat posts. In the defense, they maneuver to counterattack a flank of the enemy attack.

NOTE: Maneuver should not be done without covering fire support, but fire support elements do not have to be shooting if the enemy has not detected the maneuvering force. They must, however, always be in position to provide fire support if needed.

37

Firepower

Firepower is the capacity of a unit to place effective fires on a target. Firepower kills or suppresses the enemy in his positions, deceives the enemy, and supports maneuver. Without effective supporting fires, the infantry cannot maneuver.

Before attempting to maneuver, units must establish a base of fire. A base of fire is placed on the enemy force or position to reduce or eliminate the enemy's ability to interfere with friendly elements.

Leaders must know how to mass, control, and combine fire with maneuver. They must identify the most critical targets quickly, direct fires onto them, and ensure that the volume of fires is sufficient to keep the enemy from returning fire effectively while preventing the unit from expending ammunition needlessly.

Protection

Protection is the conservation of a unit's fighting potential so that it can be applied at the decisive time and place. Units must never allow the enemy to acquire an unexpected advantage. Platoons and squads take active and passive measures to protect themselves from surprise, observation, detection, interference, espionage, sabotage, and annoyance.

Protection includes two basic considerations: care of the soldier and his equipment, and action to counter enemy combat power.

The first consideration involves sustenance issues necessary to maintain the small unit as an effective fighting force. Such things include health, hygiene, physical conditioning, rest plans, equipment maintenance, supplies, and managing a soldiers load so that he carries only what he needs and is fit to fight when required.

The second involves security, dispersion, cover, camouflage, deception, and suppression of enemy weapons. The small unit must remain undetected to survive. Once found, the small unit becomes vulnerable to all the fires of the enemy and must either fight to break contact or close with the enemy. The small unit always wants to set the time and place of battle and must protect itself so that it can do so with maximum combat power and the important element of surprise.

Leadership

Competent and confident leadership results in effective guerrilla unit action. Good leaders give purpose, direction, and motivation in a combat situation. Only a good, effective leader can motivate his soldiers to perform difficult missions under dangerous and stressful conditions.

PLATOON OPERATIONS

There are three basic tactical operations conducted by small units: movement, offense, and defense. Small unit tactics build on the following five principles:

- Small units fight through enemy contact at the lowest possible level (i.e., without bringing other elements of the larger unit into the fight). This allows adjacent units to exploit enemy weaknesses, prevent enemy flank movements, or provide reinforcement.
- Small units in contact must establish effective suppressive fire before they or other units can maneuver. If the squad cannot move under its own fires, the platoon must lay down suppressive fires and then attempt to maneuver against the enemy position.
- Platoons and squads will fight as organized, with fire teams and squads retaining their integrity. Even buddy teams stay the same.
- Success depends on all soldiers knowing and understanding what the unit is trying to do and the specific steps necessary to accomplish the mission.

- The platoon leader never waits for the squad in contact to develop the situation. Anytime a fire team makes contact, the platoon also begins taking action. That way the platoon can quickly provide additional support, maneuver to take up the assault, or follow up on the success of the squad that made contact.

MOVEMENT

Movement refers to the shifting of forces on the battlefield. The key to moving successfully is choosing the best combination of movement techniques and formations based upon mission, enemy, terrain, troops and available time. The leader's selection of movement formations must allow squads to:

- maintain cohesion (not get separated; allow control).
- maintain momentum.
- provide maximum protection.
- make contact in a manner that allows the squad to transition smoothly to offensive or defensive action.

Formations
Formations are used for control, security, and flexibility.

Control. Every member has a standard position. Fire team leaders can see their squad leaders. Leaders control their units using hand and arm signals.

Security. Formations provide 360 degree security and allow maximum concentration of firepower to the front and flanks in anticipation of enemy contact.

Flexibility. Formations are flexible to provide for varying situations, and everyone knows the position and job of all other members.

Movement Techniques
Movement techniques describe the position of squads

and fire teams in relation to each other during movement. Platoons and squads use three movement techniques: traveling, traveling overwatch, and bounding overwatch. Considerations for planning and conducting movements to contact include:

- making enemy contact with the smallest unit possible. This allows the leaders to establish a base of fire, initiate suppressive fires, and attempt to maneuver without first having to disengage or be reinforced.
- preventing detection of units not in contact until they are in the assault.
- maintaining 360 degree security at all times.
- reporting all information quickly and accurately.
- maintaining contact once it is gained.
- generating combat power rapidly upon contact.
- fighting through at the lowest possible level.

TROOP LEADING PROCEDURE

Troop leading is the process a leader goes through to prepare his unit to accomplish a tactical mission. The procedure comprises the following eight steps:

- Receive the mission.
- Issue a warning order.
- Make a tentative plan.
- Start necessary movement.
- Reconnoiter.
- Complete the plan.
- Issue the complete order.
- Supervise.

Receive the Mission
The leader may receive the mission in a warning order, an

41

operation order, or a fragmentary order. He immediately begins to analyze it using the factors of METT-T:

What is the MISSION?

What is known about the ENEMY?

How will TERRAIN and weather affect the operation?

What TROOPS are available?

How much TIME is available?

The leader should use no more than one third of the available time for his own planning and for issuing his operation order. The remaining two thirds is for subordinates to plan and prepare for the operation. Leaders should also consider other factors such as available daylight and travel time to and from rehearsals. In the offense, the leader has one third of the time from receiving the mission until the unit begins movement. In the defense, he has one third the time from receiving the mission to the time the unit must be prepared to defend.

In scheduling preparation activities, the leader should work backward from departure time or defend time. This is reverse planning. He must allow enough time for the completion of each task.

Issue a Warning Order

The leader provides initial instructions in the warning order. The warning order contains enough information to begin preparation as soon as possible. Unit SOPs (Standing Operating Procedures) should indicate who will attend all warning orders and actions they should take after receiving it (for example, drawing ammunition, rations, and water, and checking communications equipment).

The warning order has no specific format. One possible format is provided below. The leader issues the warning order with all of the information he has at the time. He provides updates as necessary. The leader never waits for information to fill a format.

Warning Order Sample

Situation. Brief statement of enemy and friendly situation.

Mission. State in a clear, concise manner and tone. Tailor to fit the patrol; however, keep it as close as possible to the mission given in the briefing. Include: who, what, where, when, and why.

General Instructions. General instructions cover the following:

- Chain of command.
- General and special organization, to include element and team organization, individual duties, and unit equipment.
- Uniform and equipment common to all, to include identification and camouflage measures.
- Weapons, ammunition, and equipment each member will carry.
- Time schedule organized to show when, where, what, and who.
- Time, place, uniform, and equipment for receiving the patrol order.
- Times and places for inspections and rehearsals.

Specific Instructions. Specific instructions are given to subordinate leaders and to special purpose teams and key individuals who will be coordinating, performing reconnaissance, drawing supplies and equipment, preparing terrain models, etc.

An example warning order for a platoon is given below.

"This is a warning order. Hold your questions until I finish.

"The scouts have identified a squad-size enemy outpost with at least one RPG digging in to defend a bridge at grid coordinates GL126456. The rest of the enemy platoon is further west, around Hill 242.

"The captain just issued a warning order for the company to attack the squad defending the bridge and set up an ambush in case the rest of the enemy platoon attempts to reinforce the unit at the bridge or counterattack. Once the bridge is secure, it will be destroyed by explosives.

"A three-man demolitions team will be attached to our platoon.

"Our platoon, first platoon, attacks 11 July 0200 to destroy the enemy squad, seize the bridge, establish a defensive perimeter, and destroy the bridge.

"The rest of the company will depart before us. First squad second platoon will take up a supporting fire position on Hill 375 with two heavy machine guns and one mortar. During the assault, their fires will be directed by our platoon. The rest of the company will establish an area ambush along the most likely avenues of enemy approach.

"At my command, the supporting squad on Hill 375 will start suppressive fires on the enemy squad. Supporting fires will be shifted upon my command.

"You will receive more information during the patrol order.

"The time schedule is as follows. We attack at 0200 hours. The earliest we have to move is 2300 hours. My final inspection will be at 2230 hours at this location.

"We have a company rehearsal for team leader on up at 1600 hours. We will meet here at 1530 and move together to the captain's command post. I want a platoon rehearsal for team leaders, squad leaders, the medic, the forward observer, and platoon sergeant here at 1330. We will do a full platoon rehearsal at 2100 so we can do it at least once in the dark. Platoon rehearsals will be for actions at the objective. Squads rehearse breaching and react-to-contact drills on your own.

"My operations order will be here at 1030.

"I want the platoon sergeant to talk to me about resupply after this warning order. I want the platoon sergeant to plan for casualty evacuation and give paragraph 4 of the operations order.

"First squad will be the lead squad. Do a map recon prior to the operations order.

"I want the forward observer to develop a plan for fire support at the objective and discuss it with me as soon as possible after discussing it with the company operations officer.

"Second squad begin making a terrain model of the objective in 20 minutes.

"Each squad will carry an RPG to use in the assault.

"The standing signal instructions we have are still in effect.

"The time is now 0620. What are your questions?"

Make a Tentative Plan

The leader develops an estimate of the situation to use as a basis for his tentative plan. The estimate is the military decision-making process. It consists of five steps:

- Detailed mission analysis.
- Situation analysis and course of action development.
- Analysis of each course of action.
- Comparison of each course of action.
- Decision, which represents the tentative plan.

The leader updates the estimate continuously and refines his plan accordingly. He uses this plan as the starting point for coordination, reconnaissance, task organization (if required), and movement instructions. He works through this problem-solving sequence in as much detail as available time allows. As a basis of his estimate, the leader considers:

Mission

The leader considers the mission given to him by his commander. He analyzes it in light of the commander's intent two command levels higher and derives the essential tasks his unit must perform in order to accomplish the mission.

Enemy

The leader considers the type, size, organization, tactics,

and equipment of the enemy he expects to encounter. He identifies their greatest threat to his mission and their greatest vulnerability.

Terrain

The leader considers the effect of terrain and weather on enemy and friendly forces using the following guidelines:

Observation and Fields of Fire. The leader considers ground that offers him observation of the enemy throughout his area of operation. He considers fields of fire in terms of the weapons available to him (for example, maximum effective range, the requirement for grazing fire, and preplanned targets for indirect weapons).

Cover and Concealment. The leader looks for terrain that will protect him from direct and indirect fires (cover) and from aerial and ground observation (concealment).

Obstacles. In the attack, the leader considers the effect of restrictive terrain on his units' ability to maneuver. In the defense, he considers how he will tie in his obstacles to the terrain to disrupt, turn, fix, or block an enemy force and protect his own forces from enemy assault.

Key Terrain. Key terrain is any locality or area whose seizure or retention provides a marked advantage to either combatant. The leader considers key terrain in his selection of objectives, support positions, and routes in the offense, and on positioning his unit in the defense.

Avenues of Approach. An avenue of approach is an air or ground route of an attacking force of a given size leading to its objective or key terrain in its path. In the offense, the leader identifies the avenue of approach that gives him the greatest protection and places him at the enemy's most vulnerable spot. In the defense, the leader positions his key weapons along the avenue of approach most likely to be used by the enemy.

Weather. In considering the weather, the leader is most interested in visibility and the ability to move.

Troops Available

The leader considers the strength of subordinate units, the characteristics of his weapons systems, and the capabilities of attached units as he assigns tasks to subordinate units.

Time Available

The leader refines his allocation of time based on the tentative plan and any changes to the situation.

Start Necessary Movement

The unit may begin movement while the leader is still planning or forward reconnoitering. The second in command may bring the unit forward to a predetermined position.

Reconnoiter

If time allows, the leader makes a personal reconnaissance to verify his terrain analysis, adjust his plan, confirm the usability of routes, and time any critical movements. When time does not allow, the leader must make a map reconnaissance. He must consider the risk inherent in conducting a recon close to enemy forces. Sometimes the leader must rely on others (for example, scouts) to conduct the recon if the risk of enemy contact is high.

NOTE: In the insurgent environment, this information may be tentatively acquired by locals. However, when possible, at least two independent sources should be used. Check with friendly intelligence officers for information.

Complete the Plan

The leader completes his plan based upon the recon and any changes in the situation. He should review his mission as he received it from his commander to ensure that his plan meets the requirements of the mission and stays within the framework of the commander's intent.

Issue the Complete Order

Platoon and squad leaders normally issue oral operations orders. Leaders should issue the order within the sight of the objective or on defensive terrain. When this is not possible, a terrain model or sketch should be used.

NOTE: In a heavily forested (jungle) area, it is rare that orders can be issued within sight of the objective when it is an attack.

Leaders must ensure that all soldiers understand the mission and the commander's intent. Leaders ask questions to ensure that everyone understands.

Operation Order

An operation order (often called a "Five Paragraph Field Order," noted by Roman numerals below) is a directive issued by the leader to his subordinates in order to effect the coordinated execution of a specific operation.

Task Organization. Explain how the unit is organized for the operation. If there is no change to the previous task organization, indicate "no change."

I. *Situation.* Provide information essential to the subordinate leaders' understanding of the situation. This includes:

- Enemy forces. Refer to the overlay sketch. Include pertinent intelligence provided by higher HQ and other facts and assumptions about the enemy. This analysis is stated as conclusions and addresses:
    (1) disposition, composition, and strength
    (2) capabilities (a listing of what the enemy is
        able to do and how well)
    (3) most probable course of action
- Friendly forces. Provide information that subordinates need to accomplish their tasks. This covers:
    (1) a verbatim statement of the higher unit

commander's mission statement and
concept of the operation statement
(2) left unit's mission
(3) right unit's mission
(4) forward unit's mission
(5) reserve unit's mission
(6) mission of units in support or reinforcing
the higher unit

- Attachments. Attachments cover special personnel or teams attached for the execution of the mission that are not listed under task organization. List here or in an annex units all attached or detached from the platoon, together with the effective times of their attachment and detachment.

II. *Mission*. Provide a clear, concise statement of the task to be accomplished and the purpose for doing it (who, what, when, where, and why). The leader derives the mission from his mission analysis.

III. *Execution*. Give the stated vision that defines the purpose of the operation and the relationship between the enemy and the terrain.

- Concept of the operation. Refer to the overlay and concept sketch. Explain in general terms how the platoon as a whole will accomplish the mission. Identify the most important task for the platoon and any other essential tasks. If applicable, designate the decisive point, form of maneuver of defensive techniques, and any other significant factors or principles. Limit this paragraph to six sentences.
  (1) Maneuver. Address all squads and attachments by name, giving each team an essential task. Designate the platoon's main effort; that is, who

will accomplish the most important task.
All other tasks must relate to the main
effort. Give mission statements for each
subordinate element.

    (2) Fires. Refer to the fire support overlay and target
list. Describe the concept of fire support to
synchronize and compliment the scheme
of maneuver. If applicable, address priority
of fires, priority targets, and any restrictive
control measures on the use of fires.

- Tasks to maneuver units. Specify tasks, other than those listed elsewhere and the purpose of each, for squads and attachments. List each in separate numbered sub-paragraphs. Address the reserve last. State any priority or sequence.
- Tasks to combat support units. A platoon may receive an attachment of combat support units, for example, a mortar crew or heavy machine gun crew. List tasks to the combat support units in subparagraphs in the order they appear in the task organization. List only those specific tasks that must be accomplished by these units that are not specified elsewhere.
- Coordinating instructions. List the details of coordination and control applicable to two or more units in the platoon. Items that may be addressed include:
  (1) priority intelligence requirements, intelligence requirements, and reporting tasks.
  (2) Mission-oriented protective posture level.
  (3) Troop safety and operational exposure guidance.
  (4) Engagement and disengagement criteria and instructions.
  (5) Fire distribution and control measures.
  (6) Consolidation and reorganization instructions.
  (7) Reporting instructions (examples: checkpoints, scheduled radio contacts).

(8) Specific tasks that pertain to more than one squad or element.
(9) Rules of engagement.
(10) Order of march and other movement instructions (consider an annex).

IV. *Service and Support.* Include combat service and support instructions and arrangements supporting the operation that are of primary interest to the platoon. Include changes to established SOPs or a previously issued order. This paragraph is often prepared and issued by the platoon sergeant.

Reference the SOPs that govern the sustenance operations of the unit. Provide current and proposed resupply locations, casualty and damaged equipment collection points, and routes to and from them. Include information on all classes of supply of interest to the platoon, including, when applicable, transportation, services, maintenance, medical evacuation, personnel (replacements, enemy prisoners of war), and miscellaneous.

V. *Command and Signal.* Information regarding command includes:

- Location of the next higher unit commander.
- Location of platoon leader.
- Location of platoon sergeant.
- Chain of command.
  Information regarding signals include:
- Number combination password.
- Electronic security measures.
- Methods of communication in priority.
- Call signs.
- Signals to commence firing, shift firing, cease firing, final protective fires, withdraw, etc.
- Codes.
- Current time.

Supervise

The leader supervises the units preparation for combat by conducting rehearsals and inspections.

The leader uses rehearsals to practice essential tasks, improve performance, reveal weaknesses or problems with the plan, coordinate the actions of subordinate units and special teams (e.g., demo, security, assault), and improve soldier understanding and confidence. Rehearsals include the practice of subordinate leaders briefing their planned actions in execution sequence to the leader.

Leaders should conduct rehearsals on terrain that resembles the actual ground and in similar light conditions. The unit may begin rehearsals of battle drills and other SOP items before receiving the operation order. Once the order has been issued, it can rehearse mission-specific tasks.

Some important tasks to rehearse include:

- Actions on the objective.
- Assaulting a trench, bunker, or building.
- Actions at the assault position.
- Breaching obstacles.
- Using special weapons and equipment.
- Actions on unexpected enemy contact.

SECURITY

This section discusses techniques used by platoons and squads to provide security for themselves and for larger formations during movements and defensive operations.

Security During Movement

Security during movement includes the actions that units take to secure themselves and the tasks given to units to provide security for a larger force.

Platoons and squads enhance their own security during

movement through the use of covered and concealed terrain; the use of the appropriate formation and technique; actions taken to secure danger areas during crossing; enforcement of noise, light, and communications discipline; and the use of proper individual camouflage techniques.

Terrain

In planning a movement, a leader considers terrain that:

- Avoids obstacles.
- Provides protection from direct and indirect fires.
- Protects unit from observation.
- Avoids key terrain that may be occupied by the enemy.
- Allows freedom of maneuver.
- Avoids natural lines of drift.
- Avoids obvious terrain features.

If key terrain cannot be avoided, leaders plan to recon it before moving through. When operating as advance or flank for a larger force, platoons or squads may be tasked to occupy key terrain for a short time while the main body bypasses it.

Formations and Movement Techniques

Formations and movement techniques provide security by:

- Positioning each soldier so he can observe and fire into a specific sector that overlaps with other sectors.
- Placing a small element forward to allow the unit to make contact with only the lead element and give the remainder of the unit the freedom to maneuver, place effective fires, flank the enemy, etc.
- Providing overwatch for a portion of the unit.

During short halts, soldiers spread out and assume prone positions behind cover. They watch the same sectors they were assigned for movement. Leaders establish observation posts (OPs) and orient machine guns and antiarmor weapons along likely enemy approaches. Soldiers remain alert and keep movement to a minimum. During limited visibility, leaders incorporate night vision devices if available. Every few seconds, each soldier not only glances to the man on his left and right but to his fire team and squad leaders in case hand and arm signal commands are given.

During long halts, the unit establishes a perimeter defense. The leader ensures that the unit halts on defensible terrain.

For additional security during halts, a unit may establish an ambush along the route it just traveled.

Security in the Offense
Security in the offense includes actions taken by units to find the enemy, avoid detection, and protect the unit during the assault on the objective.

Movement to Contact
Platoons and squads execute guard or screening missions as part of a larger force in a movement to contact.

Reconnaissance Patrols
Recon patrols are conducted prior to offensive operations to find the enemy and determine his strength and dispositions.

Hasty and Deliberate Attacks
Small units use the same security for movement discussed above while moving from assembly areas to the objective. The base of fire and maneuver elements of the platoon must provide their own security while executing their specific tasks.

Base of Fire Element. The leader in charge of the base of fire element should designate soldiers on the flanks of the position to provide observation and, if necessary, fires to the flanks while the element engages the enemy on the objective. The base of fire element also provides security to its rear.

Maneuver Element. The maneuver element must secure its own flanks and rear as it assaults across the objective. Leaders should consider designating buddy teams to observe the flanks and rear. When clearing trenches, the maneuver element should be alert against local counterattacks along cleared portions of the trench behind the lead fire team. The base of fire element provides security for the maneuver element by engaging any counterattacking or reinforcing forces if it can do so without endangering the maneuver element with its own fires.

Consolidation

Units move quickly to establish security during consolidation of an objective. They do this by establishing OPs along likely approaches and by establishing overlapping sectors of fire to create all-around security.

Security in the Defense

Security in the defense includes active and passive measures taken to avoid detection or deceive the enemy and deny enemy reconnaissance elements accurate information on friendly positions.

General Security

The resistance movement must establish a three-layered security system. This system consists of:

Underground. The underground is the active intelligence organization that operates within enemy areas of control, such as large villages, towns, or cities. They provide early warning of enemy movements, troop concentrations, training, and possible intentions as well as target selection. The underground has other functions such as sabotage and assassination, but these are the ones that help the insurgent forces directly as far as defensive security is concerned.

Auxiliary. The auxiliary consists of farmers, local inhabitants, and others in the resistance who provide transportation, information, support, and early warning to the insurgent soldiers. They usually live in the outlying areas away from large enemy facilities. The auxiliary is careful to note any activity of the enemy, especially troop movements, equipment, discipline, and morale. The auxiliary sometimes augments the military forces when operations are conducted close to their living areas.

Insurgent Forces. These are the military units of the resistance. They get intelligence/support from both the underground and auxiliary. For effective defense security, they *must* have the active support of them.

The insurgent forces occupy a guerrilla base. This base is the primary facility of the resistance military arm. Its location should be held in secret and changed often. *No one* outside of the military arm of the resistance should know its location. This subject is very important and will be dealt with in detail later in this text.

Terrain

Leaders look for terrain that will protect them from enemy observation and fires but at the same time provide observation and fires into the area where they intend to destroy the enemy or defeat his attack. When necessary, leaders use defensive techniques, such as reverse slope or perimeter defense, to improve the security of the defensive position. They can plan protective obstacles to the flanks and rear of their positions and tie them in with supplementary fires. Leaders consider adjacent key terrain that threatens the security of their positions. They secure this terrain by posting OPs and by covering it with direct and indirect fires. Finally, leaders establish OPs along the most likely approaches into the position to provide early warning. Leaders must ask themselves how they would attack if they were the enemy.

NOTE: Insurgent forces *must* not stay in fixed positions near the enemy for extended periods. For the insurgent, mobility and stealth, along with accurate intelligence about the enemy, are very important. The insurgent force must not fight to hold ground for an extended period; it fights to hold ground only under special circumstances and for a very limited amount of time. The insurgent must choose when and where to fight.

Observation Posts

Any unit larger than a squad should post an OP, although squads often need to place them also. OPs are placed along likely avenues of enemy approach during the day. They are placed outside of hand grenade range but within small arms range (usually). This is so they can be withdrawn under fire support if necessary.

OPs can be placed to observe locations near the defensive position that could be used by the enemy to mass for an attack or reconnoiter. Leaders should consider placing OPs along areas where the enemy could move to the attack. OPs should never be left in place when the enemy has been detected near the units' position. OPs are *not* occupied by a single soldier; they are usually manned by a buddy team.

Listening Posts

Listening Posts (LPs) are a variation on the OP. LPs are used during periods of limited visibility such as night or rainy weather. Seldom are they placed in the same locations as OPs because the enemy is likely to use a different approach during limited visibility (especially at night since noise travels farther at night). LPs/OPs should have concealed return routes and be protected from direct enemy fires when possible.

Patrols

Units should actively patrol the area around their defensive position. These patrols should include observation of dead space (areas not under friendly observation), gaps between units, and any and all locations that could provide early detection by the enemy or civilians that could cause a security threat to the unit. The leader must ensure that all patrols not initiated by higher headquarters are coordinated with them for several reasons, such as chance friendly contact or early warning from headquarters of enemy units in the area.

Insurgent forces in general *must* actively patrol their area of operation to provide security for their forces and gain detailed knowledge of the terrain.

Passive Measures

These are actions taken by the unit that include the use of early warning devices, camouflage, movement control, noise/light discipline, radio discipline, night vision devices, etc.

Deceptive Measures

These include actions taken by the unit that mislead the enemy and induce him to do something against his interests. Examples include the use of dummy positions, supplemental wire, alternate positions, supplemental positions, and dummy radio broadcasts away from friendly positions.

Deceptive Operations

Small units may conduct deceptive operations as part of a larger force. These operations may include demonstrations, feints, displays, or ruses. Examples of deceptive operations include intensive recon of an objective *other* than the selected target to make the enemy think the location is about to be attacked, radio transmissions concentrated in an area close to a false target, or a large amount of radio transmissions issued by different members of a recon team to feint a larger friendly unit.

FORMATIONS

This section discusses formations, movement techniques, and actions during movement for small units.

Fire Team Formations

Formations are arrangements of elements of a unit in relation to each other. Leaders choose formations based on their analysis of the mission, enemy, terrain, weather, visibility, troops available, and time.

An effective leader is up front in a formation. This allows him to lead by example ("Follow me and do as I do.") All soldiers in a fire team must be able to see their leader.

Wedge

The wedge is the basic formation for the fire team. The interval between soldiers in the wedge is normally 10 meters. The wedge expands and contracts depending on the terrain and visibility. When the terrain, visibility, or other factors make control difficult, fire teams modify this formation. The normal distance between individuals is reduced so that fire team members can see their team leader and the team leader can see the squad leader. The sides of the wedge can move so close that the formation becomes a single file. Moving in less rugged terrain, where control is easier, soldiers expand the wedge to resume the original positions (fig. 4).

File

When terrain/visibility precludes the wedge, fire teams use the file formation (fig. 5).

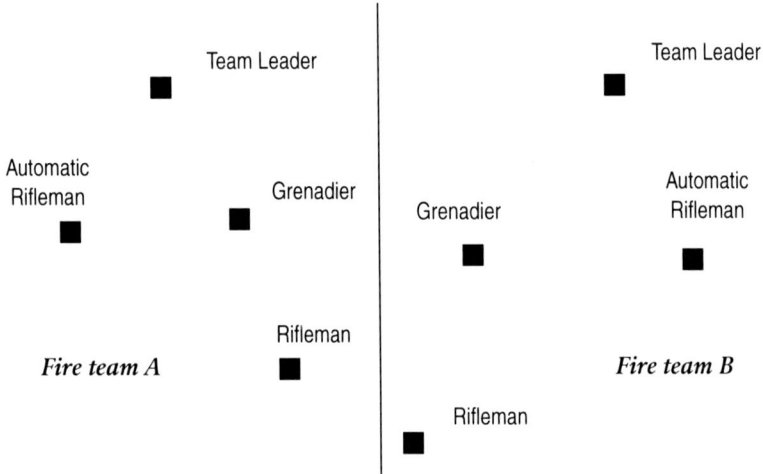

Team Leader

Automatic Rifleman

Grenadier

Team Leader

Grenadier

Automatic Rifleman

Rifleman

Fire team A

Fire team B

Rifleman

Figure 4. Fire team wedge.

Squad Formations

Squad formations describe the relationships between fire teams in the squad. They include the squad column and squad line.

Squad Column

The squad column is the squad's most common formation. It provides good dispersion from left to right and front to rear without sacrificing control and facilitates maneuver. The lead team is the base fire team (i.e., the fire team that lays down a heavy volume of fire upon contact with the enemy). When the

Team Leader

Automatic Rifleman

Grenadier

Rifleman

Figure 5. Fire team file.

60

squad moves independently or as the rear element of a platoon, the rifleman in the trailing fire team provides rear security (fig. 6).

Squad Line

The squad line provides maximum firepower to the front. When the squad is acting as the base squad, the fire team on the right is the base fire team (fig. 7).

Notice that the squad leader is positioned to control both fire teams, and the team leaders are positioned to lead by example. The automatic riflemen and grenadiers are dispersed.

Squad File

When not traveling in a column or line, squads travel in a file. If the squad leader desires to increase his control over the formation, exert greater morale presence by leading from the front, and be immediately available to make key decisions, he will move forward to

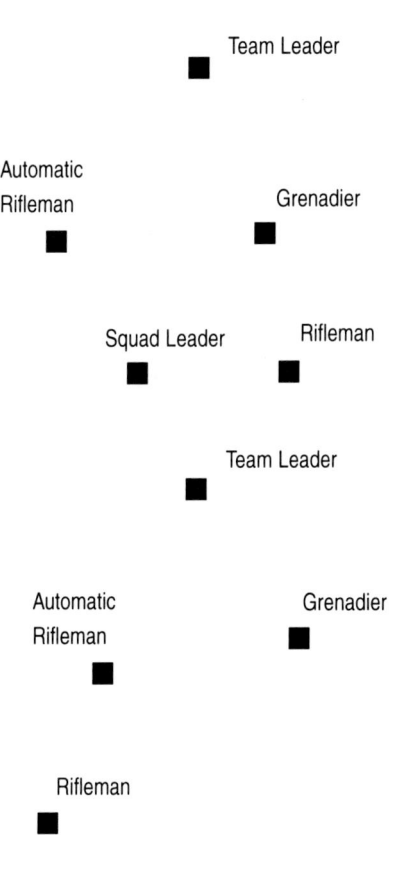

Figure 6. Squad column.

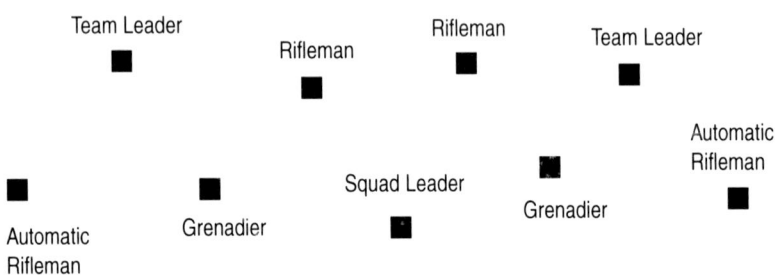

Figure 7. Squad line.

the first or second position. Additional control over the rear of the formation can be provided by moving a team leader to the last position (fig. 8).

Platoon Formations

Platoon formations include the platoon column, platoon line (squads in line or column), the platoon V, and the platoon wedge. The platoon leader should select the best formation based upon such factors as terrain, weather, enemy situation (contact anticipated or not), and visibility.

Platoon Column

This formation is the platoon's primary formation for movement. It provides

■	Team Leader
❑	Squad Leader (optional)
■	Grenadier
■	Automatic Rifleman
■	Rifleman
■	Squad Leader (normal position)
■	Team Leader
■	Grenadier
■	Automatic Rifleman
❑	Team Leader (optional)
■	Rifleman

Figure 8. Squad file.

62

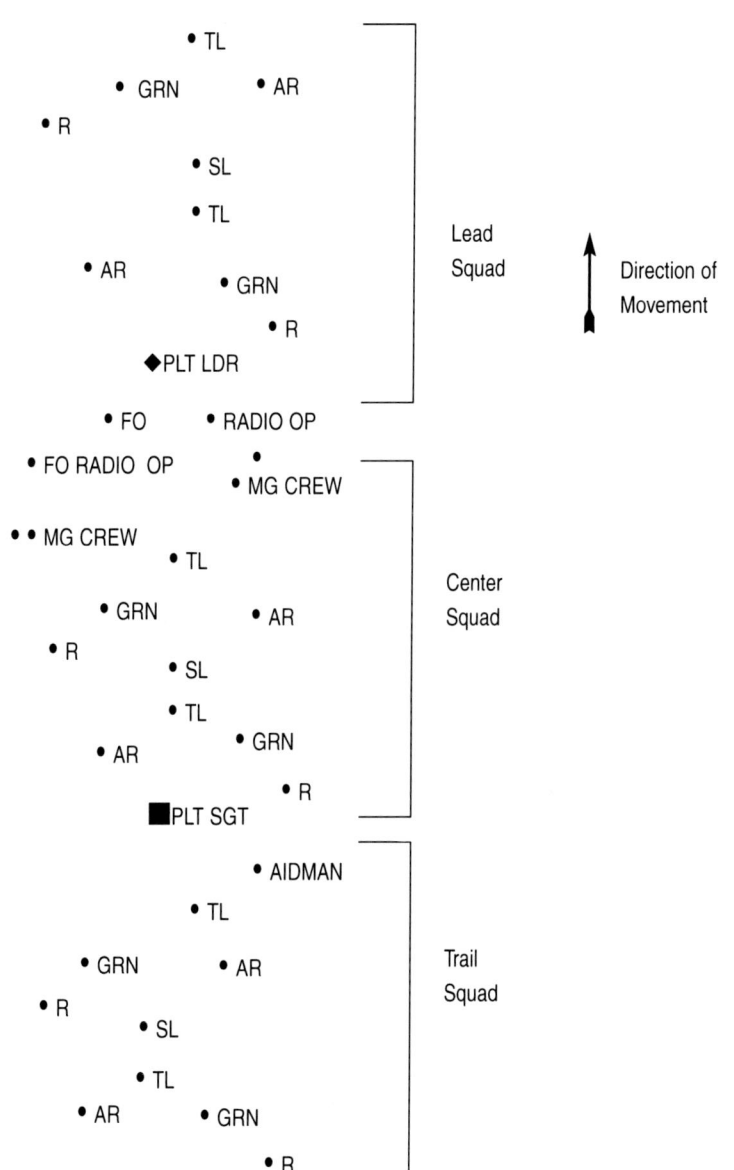

Figure 9. Platoon column.

good dispersion from left to right as well as front to rear and simplifies control. The lead squad is the base squad (fig. 9).

Notice that the platoon column allows limited firepower to the front and rear and high volume to the flanks. This is good for movement.

NOTE: The situation will dictate where the crew-served weapons move in formation. They normally will move with the platoon leader so he can establish a base of fire quickly.

Platoon Line

This formation allows maximum firepower to the front but little to the flanks. It is hard to control and does not lend itself to rapid movement. The machine guns can move with the platoon or they can support from a position. This is the basic platoon assault formation (fig. 10).

NOTE: Platoon leader positions himself where he can best control the squad.

Platoon V

This formation has two squads up front to provide a heavy volume of fire on contact. It also has one squad in the rear that can either overwatch or trail the other squads (fig. 11).

Platoon File

This formation may be set up in several methods. One method is shown below. It is used when visibility is poor due to terrain, vegetation, or light conditions. The distance between soldiers is less than normal to allow communication by passing messages up and down the file (fig. 12).

Techniques of Movement

Squad traveling is used when contact with the enemy is not likely and speed is needed (fig. 13).

Traveling overwatch is used when contact with the

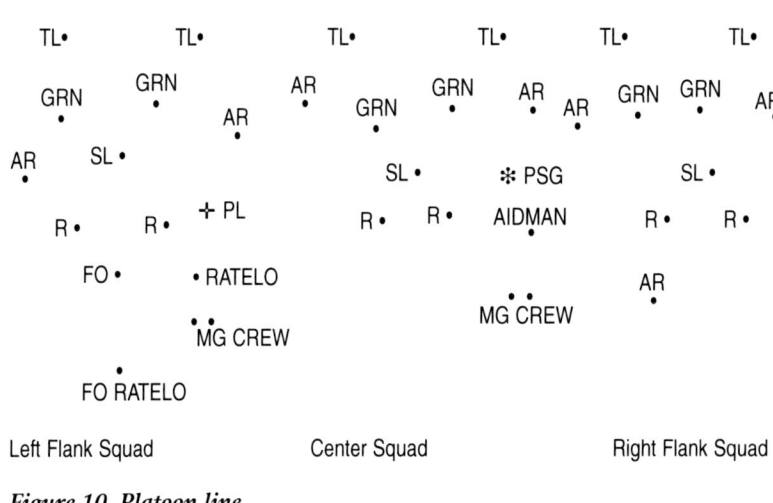

Left Flank Squad   Center Squad   Right Flank Squad

Figure 10. Platoon line.

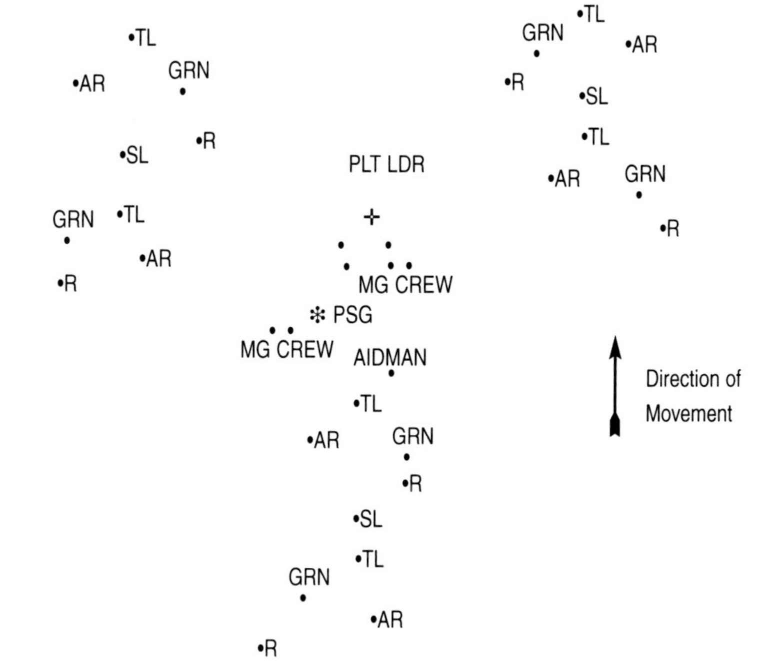

Direction of
Movement

Figure 11. Platoon V.

65

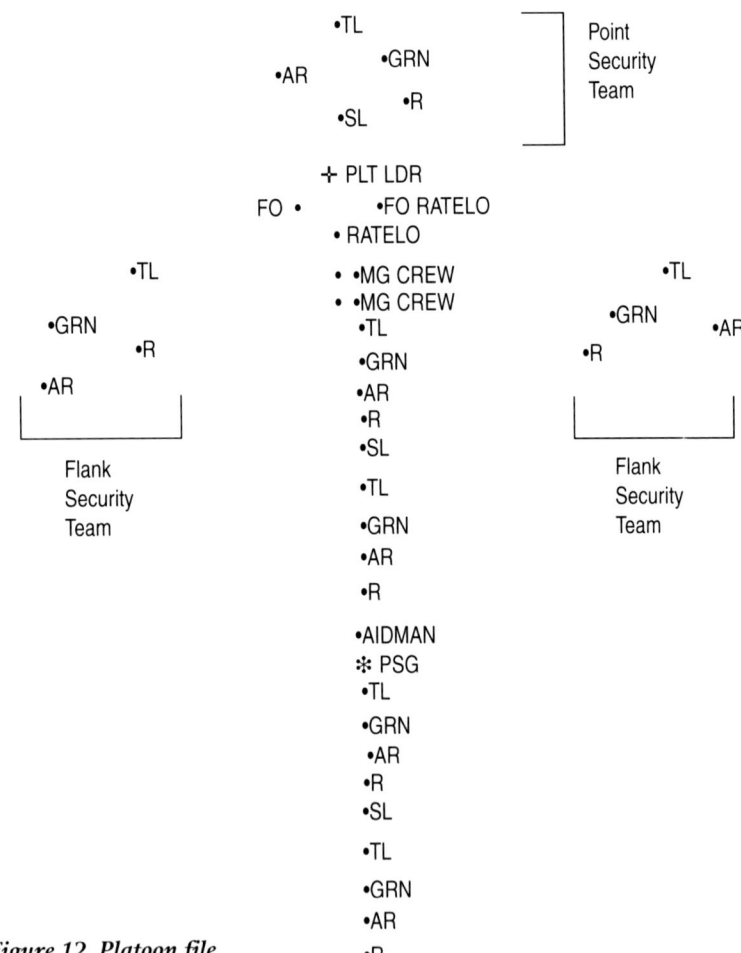

Figure 12. Platoon file.

enemy is possible. Attached weapons move near the squad leader and under his control so he can employ them quickly (fig. 14).

Bounding overwatch is used when enemy contact is expected, when the leader feels the enemy is near, or when crossing large, open danger areas.

The lead fire team overwatches first. Soldiers scan for enemy positions. The leader usually stays with the overwatch element.

The trail team bounds and signals the leader when it completes its bound and is in position to overwatch the movement of the other team.

Both team leaders must know if successive or alternate bounds will be used and which team the squad leader will be with. The overwatching team must know the route and destination of the bounding team.

Individual Movement Techniques

The following techniques are used by individual members of a small unit while on patrol or during action.

Crawl

The crawl is slow, but

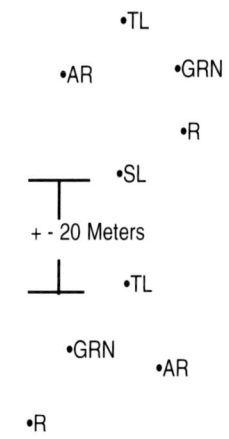

Figure 13. Squad Traveling

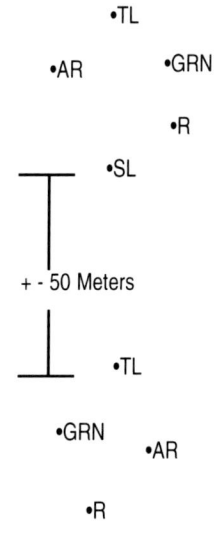

Figure 14. Traveling overwatch.

67

it helps avoid exposure to fire. If necessary, a squad will crawl all the way through its objective.

During a high crawl, the rifle is cradled over the arms and the elbows and knees/toes are used to crawl. The more body contact that is maintained to ground, the more potential for noise during movement. To reduce noise, the soldier lifts his body off the ground and crawls with his weight on his elbows and knees. The soldier ensures his weapon never touches the ground.

The low crawl is used to lower the soldier's silhouette more than the high crawl. The weapon is grasped by the sling and forward sling swivel and rests on top of one arm.

Short Rush

Short rushes from cover to cover may be used when enemy fire allows brief exposure. Men rush singly, in pairs, or by fire teams in 3- to 5-second rushes. A rush is kept short to keep enemy machine gunners from tracking rushing men. Men should not hit the ground in the open just because they have been up for 5 seconds. They must look for cover before starting the rush and then head straight to it.

Single Rush

At times, a whole platoon may have to assault an enemy position in a single, quick rush. This is done only when a platoon is under heavy indirect fire, there is no cover, a platoon is being hit by hand grenades, or the enemy that could shoot at it is suppressed.

This type of rush must be fast, and it must be accompanied by suppressive fire. A rush must be for a distance so short that the enemy can be overrun quickly.

Tactical Walk

On patrol, each soldier covers his sector of responsibility, weapon at ready. Everywhere he looks, he points his weapon.

This is done to reduce the time required to respond to chance enemy contact. Often an extended sling can be used to carry the weapon so as to reduce fatigue.

Night Movement Techniques

At night or when visibility is poor, a unit may have to move to keep pressure on the enemy, preserve secrecy, surprise the enemy, or deny him time to reorganize.

Movement techniques depend on the ability to see and control the unit, as well as on the likelihood of contact. Some ways to control a force when visibility is poor are:

- Reduce intervals between men and between units to make sure they see each other.
- Move leaders farther to the front.
- Slow down.
- Use night vision devices (if available).
- Use two small strips of luminous tape on the rear of head gear so each man can see the man in front of him. To navigate effectively at night:
- Use compass to maintain direction.
- Use a paceman to measure distance traveled.
- Choose routes parallel to identifiable terrain features.
- Move from one landmark to another (which have been picked during planning).
- Use guides or marked routes when possible.
- Have an artillery round fired on a known terrain feature.
- Have SOPs for sound and visual signals. To maintain secrecy while moving at night:
- Mask sounds of movement with artillery rounds, rain, wind, thunder, aircraft noise, etc.
- Allow no smoking, lights, and unnecessary noise.
- Use radio listening silence.
- Camouflage men and equipment.
- Use smoke to hide the unit.

69

- Use terrain to avoid detection by enemy surveillance or night observation devices.
- Take frequent listening halts.

Hand and Arm Signals

All small units must establish standard hand and arm signals. The entire resistance should attempt to standardize on the same signals. Some of the signals to be considered are:

- Halt.
- Danger area.
- Freeze.
- Hasty ambush.
- Rally point.
- Change formation.
- Enemy sighted.
- Leader forward.
- Designation of individuals to move.
- Form perimeter.
- All clear.

ACTIONS AT DANGER AREAS

A danger area is any place along a route that the patrol leader feels the unit may be exposed to enemy observation, fire, or both. Units try to avoid danger areas, but negotiating them is sometimes required in the conduct of a mission. In general, crossing danger areas is accomplished by:

- Designating near and far side rally points.
- Securing the near side (right/left sides and rear security).
- Reconnoitering and securing the far side.
- Crossing the danger area.

A recon element crosses the danger area and checks the

far side. When cleared, the recon element gives the signal. The rest of the unit crosses the danger area as quickly as possible. Rear security removes tracks.

Open Areas

Conceal the unit on the near side and observe the area. Post security to give early warning of enemy presence. Send an element across to clear the far side. When cleared, have the remainder of the unit cross at the shortest possible exposed distance as quickly as possible. If time permits, it is often better to avoid large open areas or skirt them inside the woodline. Overwatch may be used.

Roads and Trails

Cross roads or trails at a narrow spot, in a bend, or on low ground. The unit must halt on the near side and establish local security before crossing.

Villages

Try to avoid villages. Give the area a wide birth. Pass downwind to help prevent animals from getting your scent.

Enemy Positions

Pass downwind. Be alert for trip wires and warning devices.

Streams

Select a narrow spot that offers concealment on both banks. Establish local security. Observe the far side carefully. Clear the far side and establish security on the far side. Cross rapidly and quietly.

Example of a platoon crossing a linear danger area:

- When the lead element signals "danger area" (relayed throughout the platoon), the platoon halts.
- The platoon leader moves forward, confirms the danger

area, and determines the technique the platoon will use to cross. The platoon sergeant moves forward to the platoon leader.

- The platoon leader informs the squad leaders of the situation and the near and far side rally points.
- The platoon sergeant directs positioning of the near side security (usually the last squad in the formation). These two security teams may follow him forward when a danger area signal is passed back.
- The platoon leader recons the danger area and selects the crossing point that offers the best cover and concealment.
- Near side security observes the flanks and overwatches the crossing.
- When the near side security is in place, the platoon leader directs the far side security to cross and clear the far side.
- The far side security team leader establishes an OP forward of the cleared area.
- The far side security team signals to their squad leader that the area is clear. The squad leader signals to the platoon leader that the area is clear.
- The platoon quickly and quietly crosses the danger area.
- Once across the danger area, the main body begins moving slowly on the required azimuth.
- The near side security element, controlled by the platoon sergeant, crosses the danger area where the platoon crossed. They may attempt to remove any tracks left by the platoon.
- The platoon sergeant ensures that everyone has crossed and sends up the report.
- The platoon leader ensures accountability and resumes movement at normal speed.
- The platoon leader considers a change of direction in case they were seen.

PATROL PLANNING

Planning is the most crucial part of successfully accomplishing the varied tasks in patrolling.

Organization

To accomplish the patrolling mission, a small unit must perform specific tasks. A leader must determine the tasks his unit must perform and assign each and every task he has identified to the elements of his unit. Subelements maintain their integrity as much as possible down to the buddy team level. The following elements are common to all patrols:

- Headquarters. Consists of the unit leader, the radio operator (if platoon-sized unit), forward observer, a platoon sergeant, and perhaps any crew-served or special weapons that the leader wants to control directly.
- Aid and litter team.
- Enemy prisoner of war team.
- Surveillance team.
- En route recorder.
- Compass man.
- Pace man.

Initial Planning and Coordination

Leaders plan and prepare for patrols using the troop leading procedure and their estimate of the situation. Leaders identify required actions on the objective, then plan backward. The leader normally receives the mission from the headquarters S3 (operations officer) or headquarters S2 (intel officer). Coordination with higher headquarters and other units is important. Leaders normally coordinate directly with the headquarters staff and they coordinate continuously throughout the planning and preparation phases. They use checklists to prevent omitting any critical items.

Troop leading is the process a leader goes through to prepare his unit to accomplish the tactical mission. The leader goes through the following steps when leading a patrol:

- Receive the mission.
- Issue a warning order.
- Make a tentative plan.
- Start necessary movement.
- Reconnoiter.
- Complete the plan.
- Issue the complete order.
- Supervise.
  Items coordinated between the leader and the headquarters staff include:
- Changes or updates in the enemy situation.
- Best use of terrain for routes, rally points, and patrol bases.
- Light and weather data.
- Changes in the friendly situation.
- Attachments of personnel with special skills.
- Departure and reentry of friendly areas.
- Fire support available en route and on the objective and alternate routes.
- Rehearsal areas, times, and security.
- Special equipment requirements.
- Transportation.
- Signal plan (call signs, frequencies, code words, challenge, and password).
  The patrol leader coordinates with:
- The unit through which the patrol will pass and return through friendly positions.
- Other leaders that will be operating in the areas adjacent to his area.

- Other patrol leaders that have operated in the objective area or along the routes the patrol will travel.

Completion of the Plan
As the patrol leader completes his plan he considers the following:

- Essential and supporting tasks.
- Key travel and execution times.
- Primary and alternate routes.
- Signals.
- Challenge and password forward of friendly positions.
- Location of leaders during various phases of the patrol.
- Actions on enemy contact.
- Contingency plan.

Departure and Travel Through Friendly Areas
This issue requires special consideration and planning. In the guerrilla warfare environment, there are no fixed lines of battle, and there are no rear areas as such. We can consider, for planning purposes, that anytime the patrol moves through or near a friendly area, support can possibly be obtained.

The patrol leader must consider the chances of compromise before any contact is made with anyone outside the patrol. The patrol may be making contact with persons or units while en route to the objective. Valuable information may be gained from these units, and they may be able to provide guides or fire support if needed. Therefore, methods of recognition, authentication, communication, and link-up must be planned for. The mission and security of the patrol is given first priority.

Rally Points
The leader plans for methods of designating rally points

en route. Whenever possible, he physically recons the routes to select rally points. If he can only make a map recon, he selects tentative rally points.

Use of Rally Points

If dispersed before departing the friendly area, the patrol rallies at the initial rallying point. If dispersed after departing the friendly area but before reaching the first rally point en route, the patrol rallies at the initial rallying point or at the first rallying point en route. The course of action to be followed is based on careful consideration of all circumstances. The rally point to be used must be stated in the patrol order.

If dispersed between rally points en route, the patrol rallies at the last rallying point or at the next selected rally point. As before, the course of action to be followed is based on careful consideration of all circumstances. The decision is announced at each rallying point.

Ideal Characteristics of Rally Points

Rally points should:

Be easy to find. The patrol members may be occupying the rally point at night, in heavy vegetation, or during poor visibility. It is important for the patrol to consolidate and continue the mission as quickly as possible.

Have cover and concealment. The patrol may occupy the rally point after contact with the enemy, so it should provide a relatively secure location. The patrol may have to defend themselves in the rally point.

Be away from natural lines of drift. The rally point should not be in a place where people or animals tend to walk through.

Be defendable for short periods. Rally points should provide for natural defensive positions such as rocks, large trees, holes in the ground, heavy vegetation, and good observation and fields of fire. It should provide natural obstacles to enemy assault.

Types of Rally Points

Initial Rally Point. This is the first designated rally point after start of the patrol. It is located in a friendly area. It should be used as a security halt. The patrol stops and remains quiet, getting accustomed to the sounds of the surroundings. It is used to rally troops if they are dispersed before departure from friendly areas.

En Route Rally Point. En route rallying points are designated by the patrol leader while moving toward or away from the objective area. Their purpose is to provide a method of reorganization, control, and security for the patrol if dispersed or after enemy contact. Planning must include instructions on which rally points to be used (i.e. the last designated rallying point or the one before that). If the last rallying point is too near the enemy contact, it should not be used.

A silent way of designating rally points is mandatory for both day and night movement. The primary ways to designate rally points are:

- Occupy for short period. The patrol leader gives the hand and arm signal used to designate rally points. The patrol then occupies and secures the rally point for a short period of time or just halts temporarily.
- Designate with arm and hand signals while passing the rally point. The patrol leader gives the correct hand and arm signal and points to the rally point while moving. Each member of the patrol passes the signal back. Each patrol member ensures the man behind him sees and understands the signal.
- Designate with hand and arm signals while moving through the rally point.

Objective Rally Point. ORPs are rally points that are close to but out of sight, sound, and small arms range of the objective. It is usually located in the direction of movement the patrol plans to take after actions at the objective. The ORP is tentative until pinpointed.

Occupation of the ORP by a squad entails the following procedures:

- Halt beyond sight, sound, and small arms range of the tentative ORP.
- Position security.
- Patrol leader issues a contingency plan to be executed if he does not return or if the patrol is detected.
- Patrol leader moves forward with the compass man and one member from each fire team.
- Position "A" team member at 12 o'clock, "B" team member at 6 o'clock. Issue them a contingency plan and return with the compass man to the squad.
- Return to the ORP with the squad.
- Position fire team "A" from 3 to 9 o'clock. Position fire team "B" from 9 to 3 o'clock.

The occupation of the ORP by a platoon follows the same principles as outlined for the squad. The only significant difference is that there are usually three squads instead of two fire teams. The platoon leader takes a member from each squad to confirm the ORP. He stations the squad members at 10, 2, and 6 o'clock.

Reentry Rally Point. The reentry rally point is located out of sight, sound, and small arms range of the friendly unit through which the platoon will return. This also means that the reentry rally point should be outside the final protective fires of the friendly unit. The platoon occupies the reentry rally point as a security perimeter.

Near and Far Side Rally Points. These rally points are on the near and far side of danger areas. If the platoon makes contact while crossing the danger area and control is lost, soldiers on either side move to the rally point nearest them. They establish security, reestablish the chain of command, determine their personnel and equipment status, and continue the mission, link up at the ORP, or complete their last instructions.

Actions at Rally Points

Actions at rallying points must be planned and rehearsed in detail. Plans for actions at rally points must provide for the continuation of the patrol as long as there is a reasonable chance to accomplish the mission.

For example, the patrol can wait until a specified portion of the men have arrived and then proceed with the mission under the senior man present. This plan could be used for a reconnaissance patrol, where one or two men may be able to accomplish the mission. Another plan could be for the patrol to wait a specified period of time, after which the senior man present determines actions to be taken based on available personnel and equipment. This could be the plan when a minimum number of men or items of equipment, or both, are essential to accomplishment of the mission.

The first person/s to enter the rally point if the unit has been dispersed establish local security and challenge all other members as they come near the rally point. This should be done very quietly.

Leader's Reconnaissance of the Objective

The leader conducts a recon of the objective from the ORP. He confirms the location of the objective and determines suitable locations for the assault or ambush as well as security and support elements. He posts surveillance, issues a five point contingency plan, and returns to the ORP.

Debriefing

Upon completion of a patrol or operation, each member of the unit must be questioned by his leader and/or the intelligence and operations officer to ensure all information is obtained. Any deficiencies are addressed and corrections are made.

RECONNAISSANCE PATROLS

Reconnaissance patrols are used to gather or confirm information. They are supposed to see and hear and not be seen or heard. Units on a recon mission make enemy contact only if detected. They attempt to break contact as soon as possible, then continue the mission if possible.

Organization

Reconnaissance patrols can be conducted by units specially organized for recon, or by fire teams, squads, or platoons. The size and organization depends upon the type of recon, terrain, and enemy situation. Small recon patrols are more difficult to detect, but if chance contact is made with the enemy, a small patrol has less firepower to defend itself.

Elements of a Recon Patrol

Recon Element. Many times the objective can be most effectively observed by a very small recon element that detaches itself from the rest of the patrol temporarily. This is never smaller than a buddy team.

Support. The recon patrol may have a supporting element with attached weapons such as machine guns or mortars that take up a position to overwatch it and provide fire support if enemy contact is made.

Security Element. A security element is sometimes desired to isolate the objective from reinforcement or prevent the recon element from being surprised and overwhelmed.

Recon and Security. When an area or zone is the objective, the patrol is often made up of teams that provide both recon and security.

Equipment

Recon patrols rely on the ability to break contact if detected by the enemy. This means being able to deliver a high volume of fire against a possibly superior enemy force long enough to gain fire superiority and effective execution of immediate action drills. Hand grenades, grenade launchers, smoke grenades, white phosphorus grenades, CS gas, and claymore mines are among the weapons that are effective in breaking contact. Easily placed booby traps are helpful in deterring enemy pursuit as well.

Types of Reconnaissance Patrols

There are three basic types of recon patrols: point, area, and zone.

Point Recon

A point recon has an objective that stands alone from other objectives and is manned by one small element. Examples include such targets as a bridge, road junction, dam, power station, or police station.

Area Recon

An area recon has as its objective a predefined amount of terrain. Often a grid coordinate is given and the commander designates a specific area around that coordinate. In an area recon, a platoon or squad uses surveillance or vantage points around the objective from which to observe it and the surrounding area. In planning for an area recon, the leader considers the following:

- Using more than one recon and security team.
- Dispatching from an ORP. Teams depart from the ORP, observe their objective for a specified period of time, then return to the ORP. Once all information is collected, it is disseminated to every soldier.
- Dropping off recon and security teams en route.

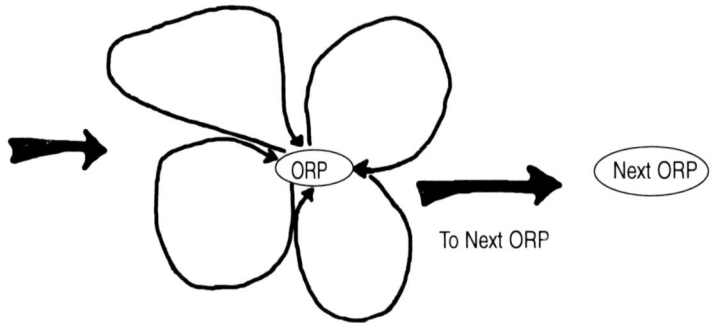

Figure 15. Fan method.

Zone Recon

A zone recon is conducted to obtain information on the enemy, terrain, and routes within a specified zone. Zone reconnaissance techniques include the use of moving elements, stationary teams, or a series of area recon actions.

When using moving elements, the leader plans the use of several squads or fire teams moving along multiple routes to cover the entire zone. Methods for planning the movement of multiple elements through a zone include the fan, converging routes, and successive sectors.

Fan. The patrol leader first selects a series of ORPs throughout the zone from which to operate. When the patrol arrives at the first ORP, it halts and establishes security. The patrol leader confirms the patrol's location. He then selects routes that form a fan-shaped pattern out from the ORP (fig. 15). Once the routes have been selected, the leader sends out recon elements along them. He does not send out all personnel at once, as he must maintain security at the ORP with one of his elements. The teams are sent out along adjacent routes so as not to make chance contact with each other. All teams move in the same direction (i.e., clockwise or counterclockwise).

Converging Routes. The patrol leader selects routes from the ORP to a link-up point at the other side of the zone. Each recon team uses the fan technique along their assigned routes through the zone. A time is set for all teams to meet at the far side of the zone (fig. 16).

Successive Sector. The patrol leader divides the zone into sectors. Recon teams use the converging routes technique to reconnoiter within each sector. Each team arrives at a designated intermediate link-up point, where it shares the information gathered to that point before moving on to reconnoitering the next sector (fig. 17).

COMBAT PATROLS

Combat patrols are conducted to destroy or capture enemy soldiers or equipment; destroy installations, facilities, or key points; or harass enemy forces. An attack can be used as a diversion or to lure enemy reinforcements into a kill zone. There are two types of combat patrols: ambushes and raids.

The leader conducts a recon of the objective from the ORP. He confirms the location of the objective and determines suitable locations for the assault or ambush and security and support elements. He posts surveillance, issues a five-point contingency plan, and returns to the ORP.

Combat patrols have the following elements or teams:

Assault. The part of the patrol that deploys close enough to the objective to permit immediate assault of the objective if detected by the enemy.

Security. Security elements isolate an objective to prevent enemy reinforcement, flanking, and counterattack movements, or enemy escape from an objective. Security elements also are used to protect ORPs and assembly areas.

Support. The support element moves into position prior to the assault element so that it can suppress the objective and shift fire when the assault starts.

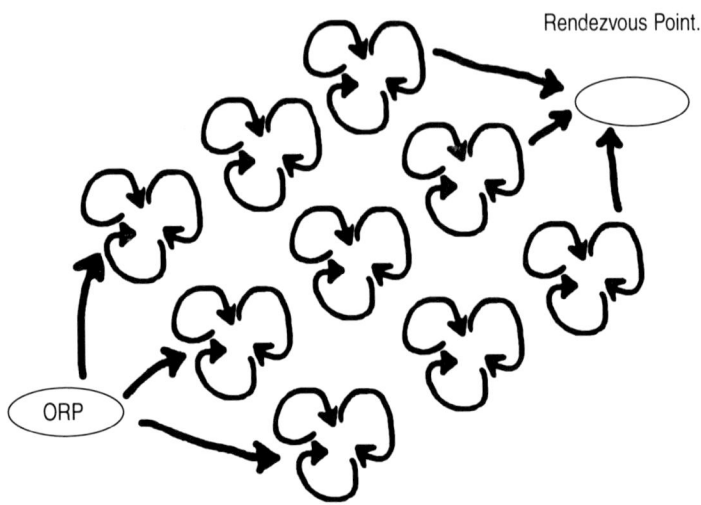

Rendezvous Point.

Figure 16. Converging routes method.

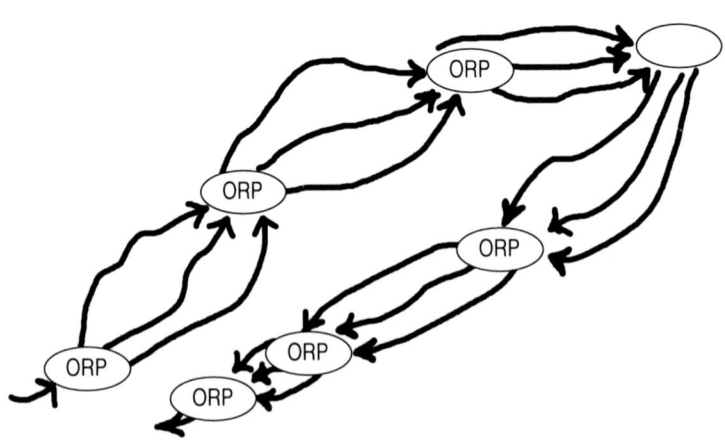

Rendezvous Point

Figure 17. Successive sector method.

Search Team. Usually part of the assault element. It searches enemy dead for documents and equipment.

Breach Team. Usually part of the assault element during the attack of a fortified enemy position. It is their task to penetrate the enemy defenses.

Demolition Team. Usually part of the assault element during the attack of a fortified enemy position or during an ambush. This team prepares charges and destroys enemy facilities and/or equipment with demolitions.

Ambush

An ambush is a surprise attack from a concealed position against a moving or temporarily halted target.

Types of Ambush

Point Ambush. A point ambush is conducted in a single isolated kill zone. The leader should consider the following sequence of actions when planning a deliberate point ambush:

- The security and surveillance teams should be positioned first.
- The support element should be in place before the assault element moves into position.
- The support element overwatches the movement of the assault element into position.
- The patrol leader is in charge of the assault element. He must check each soldier after they establish their position. He signals the surveillance team to rejoin the assault element.

Area Ambush. An area ambush is made up of more than one point ambush conducted at the same time in the same general area.

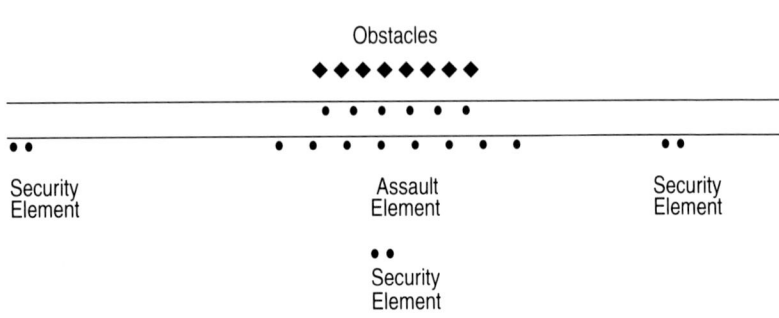

Figure 18. Linear ambush.

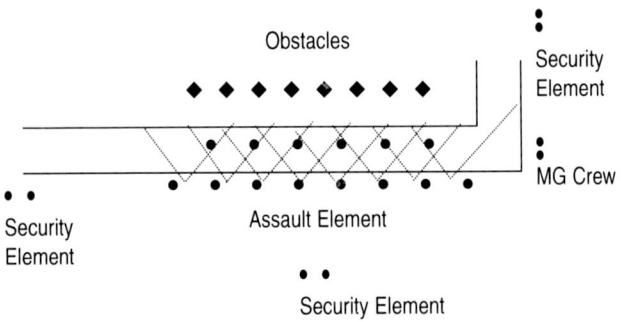

Figure 19. L-shaped ambush.

Ambush Formations

Linear. A "straight line" ambush, usually along a road or trail. The assault element lays in wait along the road for the enemy (fig. 18).

L shaped. This formation has a leg at a right angle to the rest of the assault element. It is a little more difficult to control than a linear ambush, but it is well worth the trouble (fig. 19).

Assault. The assault element fires into the kill zone and may assault it. The following steps make up the assault portion of an ambush.

86

- Establish individual sectors of fire as assigned by the patrol leader.
- Emplace aiming stakes.
- Emplace claymores and other protective devices.
- Emplace claymores, mines, or other explosive in dead space within the kill zone.
- Camouflage positions.
- Take weapons off safe. The sound of moving the selector switch while the enemy is near the kill zone could compromise the ambush. This should be the last action taken before waiting for the enemy to approach the kill zone.
- Patrol leader initiates the ambush. A claymore may be used. A backup must also be planned for initiating the ambush. This should be a casualty-producing weapon such as a machine gun.
- The initiation and initial volley of fire should be violently devastating.
- Mark any enemy equipment to be carried back. Captured weapons are cleared and put on safe.
- Prepare nonessential equipment for destruction with a dual-primed charge.
- Demo team awaits for the signal to detonate. This is normally the last action of the assault element on the objective and may be the signal for the security teams to return to the ORP.

Security. The security element isolates the objective area to prevent enemy escape or reinforcement and notifies the leader when the enemy is approaching the kill zone. It must also notify the leader if another enemy unit is following the lead unit. It also secures the ORP.

If a flank security team makes contact, it fights as long as it can without becoming engaged decisively. It uses a predefined signal to let the platoon leader know it is breaking contact. The patrol leader may designate a portion of the support element to help them break contact.

The security element secures and mans the ORP during the operation. Elements normally withdraw in reverse order that they established their positions.

When elements return to the ORP, the security element maintains security of the ORP while the rest of the patrol prepares to leave.

Support. This element supports the assault element by firing into and/or around the kill zone, and provides supporting fires during withdrawal. The support element identifies sectors of fire for all weapons, especially machine guns. It emplaces limiting stakes to prevent firing into friendly positions when using the L-shaped ambush, and it emplaces claymores and other protective devices.

Search. On a prearranged signal, the search element moves to the kill zone and conducts a search of enemy dead. This must be accomplished quickly. The team moves from one side to the other, marking the bodies after each is searched to ensure complete coverage and prevent duplication. The search team uses the two-man search technique, where one man points his weapon at the enemy soldier in case he is still alive while the second man searches the body. Once the KIA is searched, he is placed on his back with his arms crossed to indicate that he has been searched.

Selection of an Ambush Site

Selection of an ambush site depends on the mission. The routes to and from the ambush site must provide for stealth and provide cover and concealment. Do not use the same route from the objective that was used to get to it. Change directions often by dividing the route into legs.

Factors in selecting an ambush site include:

- Probable size and composition of the enemy force that is to be ambushed.
- Likely formations the enemy may use and his reinforcement capability.
- Terrain along the route that provides unobserved avenues of approach and withdrawal.
- Timing of the ambush.
- Whether it will be a day or night ambush.
- Types of weapons to be employed in the assault.
- Fields of fire

NOTE: The terrain at the site should help funnel the enemy into the kill zone and offer natural obstacles that will restrict enemy maneuver. There should be a minimal number of possible enemy approach routes into the objective area. The terrain should offer multiple avenues of approach and withdrawal into and out of the objective area for the attacking force.

Communications and Control

Communication between the elements consists of visual, audio, and electronic. The security element must be able to report to the patrol leader the approach of the enemy. The patrol leader must be able to communicate with the support element to call for or adjust fires, and he must be able to give commands to control all elements and receive status reports from them.

Methods of control without radio communications consist of such things as flares, noises, lights, tugging on wires, whistles, explosives, tracers, and specific times for executing actions. Methods of control with radio consist of not transmitting unless absolutely necessary, use of the squelch, breaking squelch a specific number of times, transmitting codes, and use of headsets.

Control is vital yet more difficult at night. Positioning men closer helps in control at night.

Considerations of a Successful Ambush

- Cover the entire kill zone by fire.
- Use a well-trained, highly disciplined team.
- Have a simple, effective plan. Every man must know his duty.
- Recon the objective. Use caution to prevent disclosing position. Avoid crossing the kill zone.
- Establish all-around security during all phases of the operation.
- Place men and weapons carefully. Concealment is first priority.
- Use camouflage during movement to and from the objective and in the ambush site.
- Have an easily understood signal to open fire and cease fire.
- Use existing natural obstacles or place obstacles (claymores, demo) to keep the enemy in the kill zone. Examples of natural obstacles include embankments, bodies of water, and cliffs.
- Protect assault and support elements with claymores, explosives, or other obstacles.
- Employ preplanned direct and indirect fires, mines, and booby traps to inflict damage and casualties, isolate the objective to prevent reinforcement or escape, and cover the withdrawal of the assault element.
- Use search teams to search the dead enemy. They must be able to move past their own protective obstacles quickly.

Rehearsal

Ambush rehearsals are extremely important and should be conducted in an area that is as similar as possible to the

90

area of operations. Rehearsals must not be limited to actions in the objective area. Units should have established standards for certain things such as immediate action drills, actions at danger areas, basic load, etc., but that does not change the need to practice those predefined actions in order to ensure adherence.

Objective Rally Point

The ORP is a point that is out of sight, sound, and small arms range of the objective. It is usually in the direction of movement the patrol plans to take after the ambush. The ORP is tentative until pinpointed.

Actions at or from the ORP include:

- Recon of the objective
- Issuing a FRAGO (fragmentary order—a partial patrol order consisting of only those items that have changed).
- Dissemination of information after a recon if contact was not made.
- Final preparations before continuing operations, including lining up rucksacks, reapplying camouflage, preparing demolitions, preparing litters, and inspections.
- Actions after the ambush, including , accounting for soldiers and equipment, redistribution of ammunition, and reestablishing the chain of command.

Reasons for Failure
- Premature initiation of the attack. The assault element members must wait until the command for fire is given. If part of the enemy force is outside the kill zone, it is a threat to the assault element. If the assault element attacks a point element of a larger force, it can be flanked or cut off and surrounded.
- Discovery of the ambush due to poor noise and light discipline.

91

- Poor camouflage.
- Poor site selection.

Miscellaneous Ambush Tips
- It is best to move into position at last light.
- Area weapons are most effective.
- Patrol members must not sleep in ambush. If there is a danger of this due to excessive fatigue, some way of ensuring men stay awake is mandatory. For example, a cord can be attached to each man's thumb. Whenever a man receives a tug, he returns it. The cord must be easily removed and quiet.
- The patrol leader should plan for illumination of the objective if needed.
- A vehicle ambush can be organized similar to a linear ambush. Planning considerations should include stopping the leading and last vehicle in the column to prevent escape from the kill zone.
- Often small arms ammunition is in short supply in an insurgent war. In this situation, explosives are often the best means to conduct an ambush and certain other operations since they can be made from many common chemicals.

Raids
A raid is a surprise attack against an enemy force or installation. Such attacks are characterized by secret movement to the objective area; brief, violent combat; rapid disengagement from action; and swift, deceptive withdrawal.

Raids are conducted by guerrilla units to destroy or damage supplies, equipment, or installations such as command posts, communication facilities, depots, and radar sites; capture supplies, equipment, and key personnel; or cause enemy casualties. Other effects of raids are to draw attention away from other operations; keep the enemy off balance; and force him to deploy additional units to protect his rear areas and

92

facilities. Raids are sometimes used to lure a reinforcing unit into a carefully staged ambush.

Elements

Leader and Headquarters. Consists of the operations leader, radio operator, and other personnel in the command element of the operation.

Assault Team. The assault element deploys close enough to the objective to permit immediate attack if detected by the enemy. As supporting fire is lifted or shifted, the assault element attacks, seizes, and secures the objective. It protects demolition teams, search teams, and other teams while they work. On order or signal, the assault element returns to the ORP.

Demo Team. Prepares, sets, and detonates explosive charges to destroy enemy targets. It is part of the assault element.

Breach Team. Clears paths through enemy obstacles. Don't overlook the possibility of using animals (cattle, sheep, goats, etc.) to help detonate mines and booby traps, breach obstacles, and draw fire from the enemy to discover his positions. Part of the assault element.

Aid/Litter Team. Medics and stretcher bearers. Part of the assault element.

Search Team. This element searches enemy bodies for documents and other forms of information. Part of the assault element.

Reserve. This element is retained (if resources allow) so that a penetration of the enemy's defenses can be exploited more rapidly. The reserve also gives the patrol leader flexibility if one of the elements needs to be strengthened or an unexpected event takes place.

Support. The support element provides suppressive fires on the objective, at personnel, and at key areas near or beyond the objective to destroy enemy weapons. This allows the assault element to close with and destroy the enemy.

The support element can help deter enemy reinforcements, shift its fires upon signal to destroy the enemy during withdrawal, aid the assault element if it must withdraw under pressure, place smoke on the objective to limit the enemy's vision, and fire illumination rounds at night. If it is not feasible to provide support from an advantageous location, this element can accompany the assault element. The support element must provide security for its position.

Security. The security element prevents enemy reinforcements and escape and secures the ORP. Additionally, the security element covers the withdrawal of the assault element and acts as rear guard for the raid force.

Planning Considerations

Recon. Detailed information is essential to a successful raid. Knowing the location and disposition of enemy reinforcements and the routes they can take to reach the objective are important. All information concerning sentry locations and schedules, weapons emplacements, communications facilities, barracks, and terrain to/from objective command post are just some of the details required. Many times this can all be obtained by effective recon and surveillance. Local civilians can also provide such intelligence in a guerrilla warfare environment. When the leader conducts his recon of the objective prior to the attack, he posts surveillance on the objective.

Maximum Use of Surprise. Prevent detection of elements until they are in the assault. Guerrilla activities can be planned so they give no indication of the targeted objective. Often, movement during darkness or rain or under distant artillery noise can help cover the movement of the raiding party. Remember that sound travels farther at night, so every effort must be made to prevent noise by silencing equipment and observing noise discipline. Smoke can also be used to help conceal the movement of a raiding party.

Objectives. Squad leaders are given specific objectives to neutralize. These leaders direct their unit's fires against these objectives.

Routes to/from the Objective. These routes should provide stealth, cover, and concealment. Consider the use of guides from the local auxiliary. Proper formations are used en route to the objective to avoid contact and achieve surprise.

The platoon must remain undetected. If detected early, it concentrates direct and indirect fires, establishes fire superiority, and maneuvers to regain the initiative. To aid in maintaining surprise, the leader can consider arranging a dummy attack at another location, or probing attacks can be conducted over an extended period prior to an actual assault. In the guerrilla warfare environment, an unsuccessful raid is terminated to prevent undue loss.

Booby traps can be placed along the withdrawal routes from the objective to slow down any pursuers. Guides need to be used to assist the raiding party in passing through the lanes in the booby traps.

Preplanned Fires. Direct and indirect fires, mines, and booby traps can be employed to inflict damage and casualties on the objective, isolate the objective to prevent reinforcement or escape, cover the withdrawal of the assault element, help defend against counterattack, or mask the movement of the raiding party.

Use of Illumination. Illumination can be used to allow the raid to take place during darkness. The illumination should be placed behind the defenders so that they are silhouetted against the light, but the loss of night vision is a concern. Be aware that the enemy may use illumination, which would also cause a reduction in night vision.

Don't overlook the use of binoculars in night operations. They concentrate the available light.

Assault of the Objective. Special teams such as sapper, breach, demo, or sentry elimination personnel move into position and execute their tasks. Suppressive fires are placed upon the target to gain fire superiority and destroy enemy positions and weapons.

Once suppressive fires are effective, the maneuver element closes on the objective. Leaders designate specific targets for fire teams/squads to attack. As the assault element gets closer to the enemy, there is more emphasis on suppression and less on maneuver. Ultimately, all but one fire team may be suppressing to allow that one fire team to maneuver and break into the enemy position. Often, supporting fires are lifted or shifted just prior to the advance of the assault element onto the objective.

Consolidation and Reorganization. Once the objective has been overrun, the assault element sets up a defensive perimeter in case of a counterattack. Search teams search enemy dead. Demo teams set charges. The chain of command is reestablished. Ammunition is redistributed.

All-around security is critical. During the leader's recon of the objective, he considers the terrain and selects tentative defensive positions, notes key terrain from which the enemy could launch a counterattack, and considers placing preplanned fires on them to cover approaches from those points.

One technique used to set up a defensive perimeter quickly is the "clock" technique. The patrol leader designates a compass direction or direction of attack as 12 o'clock and uses clock positions to identify left and right boundaries for the squads. He then positions key weapons to cover the most likely avenues of approach. The patrol leader often designates the assistant patrol leader to take responsibility for security.

As soon as all actions on the objective are complete, the assault element withdraws quickly. It is covered by the security and sometimes the support elements over predetermined routes through a series of rally points.

Should the enemy organize a close pursuit of the assault element, the security element assists by fire and movement,

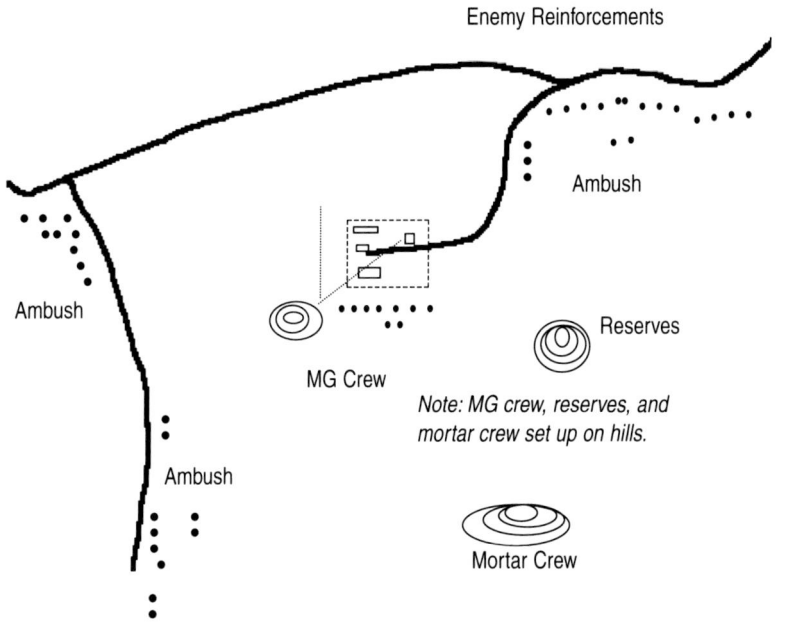

Figure 20. An example of a raid used to lure enemy into an ambush when it tries to reenforce a facility under attack.

distracting the enemy and slowing him down. Elements of the raiding party that are closely pursued by the enemy do not attempt to reach the initial rally point but, on their own initiative, lead the enemy away from the remainder of the force and attempt to lose him by evasive action over difficult terrain. If the situation permits, an attempt is made to regroup with the main body at another rally point or continue to the base as a separate group. When necessary, the pursued element breaks into small groups to evade the enemy.

Signals. Many signals may be required during the conduct of a raid. From the point the patrol reaches the ORP, signals may be required for such things as initiation of supporting fires, initiation of the assault, lifting or shifting supporting fires, concentrating fires, cease fire, withdrawal, etc.

Fire Control. Control can be a significant problem during a raid. Guerrilla units without extensive radio communications equipment will find coordination between various widespread elements difficult to achieve. Pyrotechnics, audible signals, runners, or predesignated times may be used to coordinate action. Tracers can be used by leaders to designate where to concentrate fires, or machine guns in the support element can fire tracers across the objective to mark the beginning edge of supporting fires. Tripod-mounted weapons can be trained on primary targets during daylight hours.

Rehearsals. Rehearsals are key to the success of raids. They should be conducted on terrain similar to that of the objective. Mock-ups of enemy positions are helpful. Once all tasks have been learned, they should be rehearsed during the light conditions that will exist during the raid. Terrain models should be used in briefing participants in the operation.

IMMEDIATE ACTION DRILLS

Immediate action drills are prearranged responses to common situations in combat that require instant action from a small unit. They may be offensive or defensive in nature. In order to be successful, they must be well-rehearsed and understood by all members of the unit, and they must be simple and quick.

Chance Contact

Chance contact refers to an accidental or unexpected enemy sighting. The enemy may or may not see the friendly force. The specific situation will dictate the action to be taken. There is often not enough time to give detailed instructions, so each member of the patrol must know what to do. These actions must be practiced until they are second nature.

Freeze

Pointman sees enemy but enemy does not see patrol. Pointman signals "freeze." Signal relayed back through patrol. All patrol members halt and remain motionless to allow enemy to pass. Any member seen by the enemy opens fire immediately.

NOTE: The human eye is conditioned to detect sudden movement. Therefore, the signal to freeze must not be easily detected by the enemy. This means no sudden holding up of a hand or exposing the palm of the hand toward the enemy.

Hasty Ambush

Pointman sees enemy but enemy does not see patrol. Point signals "freeze," then "enemy front" using hand and arm signals. Signals are relayed back quickly. If the patrol leader makes the decision to execute a hasty ambush, he gives the appropriate hand and arm signal and direction of movement. Enemy is allowed to proceed as far as possible into the kill zone. Patrol leader initiates ambush by firing first, aimed shot. Based on the mission, the patrol may conduct an assault into the kill zone or withdraw.

Immediate Assault

Patrol and enemy see each other at the same time. First patrol member to see the enemy quickly fires an aimed shot, then shouts, "contact front," "contact right," etc. Patrol quickly moves on line and assaults in the direction of the enemy, moving aggressively and shooting only at seen targets.

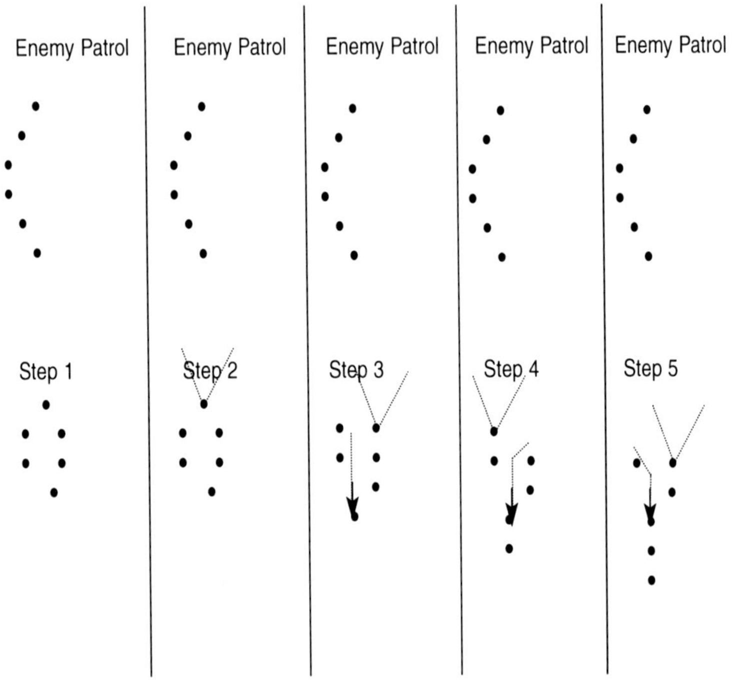

Enemy Patrol	Enemy Patrol	Enemy Patrol	Enemy Patrol	Enemy Patrol
Step 1	Step 2	Step 3	Step 4	Step 5

Figure 21. Withdrawal by fire.

Counterambush

If part of the patrol is caught in the kill zone, they return fire and move out of the kill zone quickly. Patrol members not engaged should assist the group in the kill zone as quickly as possible.

If the entire patrol is caught in the kill zone, it immediately returns fire and moves out of the kill zone. If it is a close ambush and no cover is available, the only alternative may be to attack directly into the enemy force. After the patrol breaks contact, it reorganizes at the last designated rally point.

100

Withdrawal by Fire

This technique is usually used by small units to break contact with larger enemy forces (fig. 21). If the patrol and enemy sight each other at the same time, the first person to see the enemy fires full automatic into the enemy and shouts "enemy front," "enemy right," etc. All other members in the patrol go down on one knee in a staggered pattern. The man farthest away from enemy prepares a white phosphorus grenade (if the wind is not coming from the direction of the enemy) or a high explosive grenade. As soon as the first man finishes one magazine, he goes to high port, turns, and runs away from the enemy, reloading as he moves. The second man opens fire on full auto, trying to initiate his fire at exactly the same time as the previous man expends his last round. When his magazine is emptied, he goes to high port and follows the first man. This action is repeated for every man in the patrol until the last man. The last man waits until the man before him starts firing. He shouts, "grenade out" and throws his grenade toward the enemy. He and the man firing go to high port and run. When the grenade detonates, the patrol member that threw it stops, turns on one knee, and fires full auto toward the enemy. The entire scenario is repeated if necessary (usually determined by the situation or patrol leader).

React to Artillery Fire

All members hit the ground when incoming rounds are first heard. After impact, the entire patrol moves a predetermined distance at a run in the direction of march. Patrol leader may change direction by the clock method.

React to Aerial Attack

Upon approach of enemy aircraft, patrol takes cover and allows aircraft to pass. If the aircraft attacks, patrol fires either automatically or on patrol leader's command. All patrol mem-

101

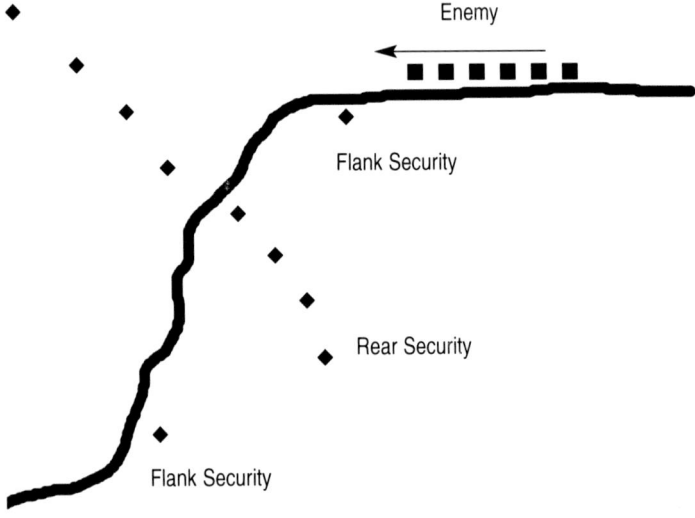

Enemy

Flank Security

Rear Security

Flank Security

Figure 22.

bers fire at full automatic ahead of aircraft so that it flies into the fires.

Actions at Halts

For short halts, the patrol establishes all-around security and covers all approaches into the area. This is best accomplished with a cigar-shaped perimeter. The assistant patrol leader moves forward through the unit, checking security as he goes, and meets the patrol leader to determine the reason for the halt and render assistance.

For long halts, the patrol occupies a security perimeter. Techniques are similar to those used for occupying a patrol base.

Examples of Immediate Action Drills

Situation: Part of the patrol has crossed the trail. The

102

enemy is sighted by flank security. The enemy has not detected the patrol (fig. 22).

- Enemy is spotted by flank security.
- Security signals "enemy sighted."
- Patrol freezes. Anyone exposed moves smoothly, quickly, and quietly to covered and concealed position.
- The patrol allows the enemy unit to pass.
- After waiting to see if the enemy unit was the lead element of a larger unit, the rest of the patrol crosses.
- Patrol changes route of march.
- After moving out of small arms range, unit conducts security halt.

Situation: Patrol is sighted by enemy on road or trail while patrol is crossing (fig. 23).

- First person to realize the enemy has detected patrol fires an aimed shoot at the enemy. Patrol member shouts, "enemy left/right, etc." Then all members take cover.
- Patrol members on near side provide a base of fire.
- Element on far side moves away from the enemy and observes the trail.
- If no enemy is sighted, the troops on the far side cross to the near side and take up an overwatch position.
- Flank security signals to the element in contact that the element is in position.
- Unit breaks contact. This may require the overwatch element to lay down a base of fire. Normally, the element in contact will assist the overwatch element in gaining fire superiority and wait until the overwatch element can maintain fire superiority before attempting to move.

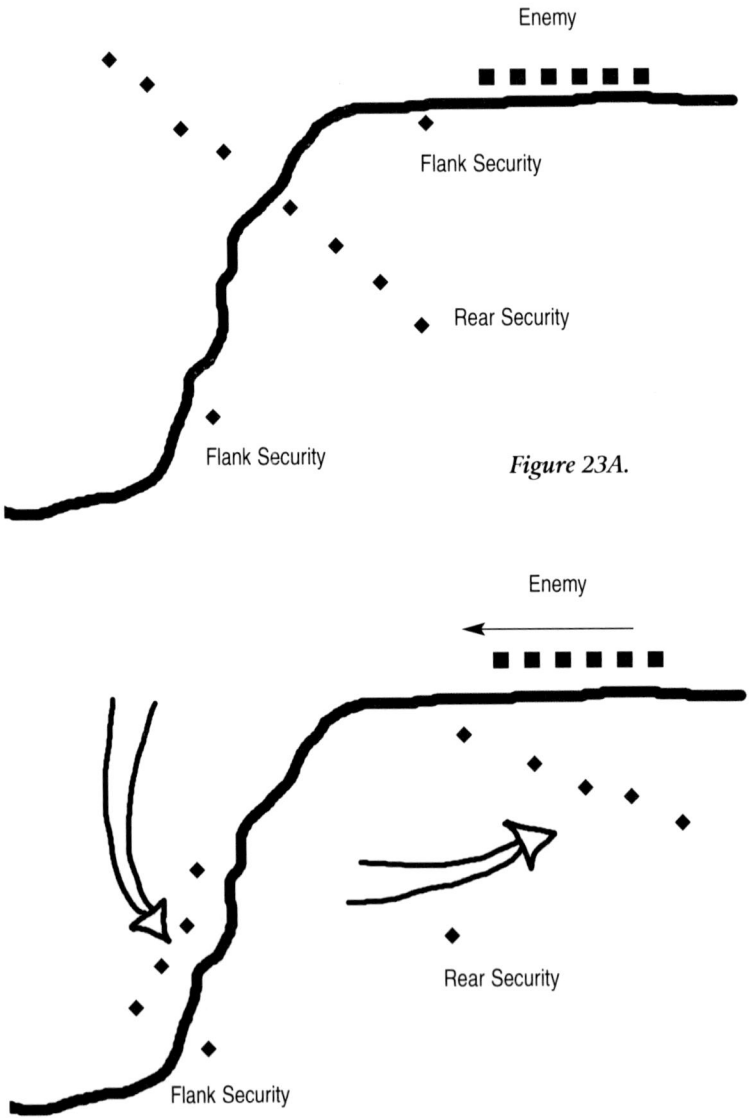

Enemy

Flank Security

Rear Security

Flank Security

Figure 23A.

Enemy

Rear Security

Flank Security

Figure 23B.

104

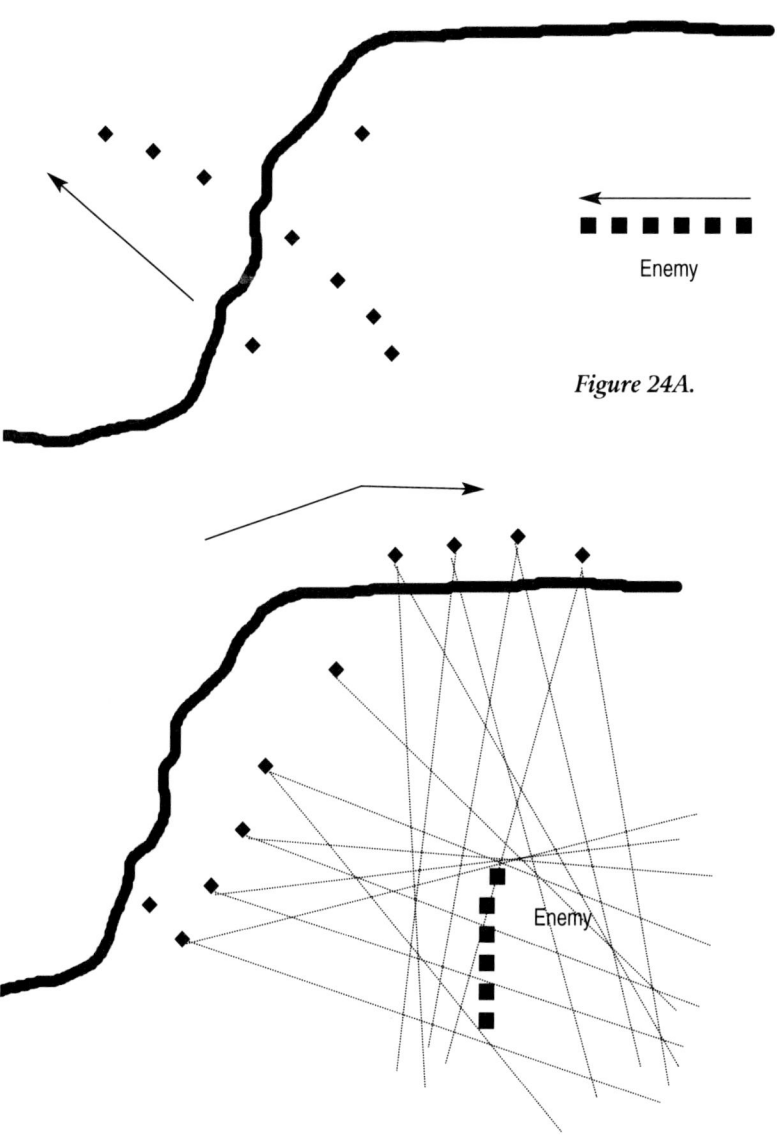

Figure 24A.

Figure 24B.

105

Situation: Patrol crossing road or trail makes contact with enemy on near side. Part of patrol is on the far side (fig. 24).

- Near side element immediately opens fire to gain fire superiority.
- Far side element maneuvers to establish a base of fire.
- Near side element breaks contact (possibly using grenades) and crosses trail using cover and concealment. The element crossing attempts to cross between flank security elements if the terrain and situation permits; however, if it is safer to go around the security, they do so, first ensuring the security element knows they will be crossing there.

Situation: Patrol has encountered a trail. Security has been established and two men sent to the far side to recon. After crossing the trail, the two men are brought under fire by the enemy (fig. 25).

- Main body takes covered positions and provides a base of fire. Rear security is maintained. If the enemy tries to flank, the security element on that side takes them under fire.
- Recon element breaks contact. They may use hand/CS grenades to break contact.
- Recon element rejoins main body.
- As soon as recon element rejoins main body, patrol leader signals to break contact.
- Patrol uses fire and maneuver/movement to break contact.
- Fire team not in contact takes up an overwatch position and establishes a base of fire. Team in contact withdraws.
- Patrol uses bounding overwatch until contact is broken. It moves to the last designated rally point, where

it redistributes ammo, reestablishes chain of command, and continues mission.

Situation: Patrol comes under artillery fire.
- Incoming rounds heard. First person to hear shouts "incoming."
- Everyone hits ground.
- After initial impact, patrol leader gives direction of movement and distance. The patrol runs in the specified direction and distance to get out of the impact area.

BREAKOUT FROM ENCIRCLEMENT

If a small patrol is surrounded by a larger force, it will most likely have to fight its way out. The technique below can serve as a guide in a unit's development of its *predetermined and rehearsed* breakout method.

When encircled, it is best to attempt a breakout as soon as possible. This is because the enemy will have less time to pinpoint your exact position, will probably have gaps in his lines, and will probably not have located key weapons, requested fire support, placed blocking forces, positioned reserves, etc. Essentially, the sooner the breakout is attempted, the better the chance of success with fewer casualties.

Preparation for Breakout
Plans must be made prior to the attempt to take care of the following:

- Rucksacks and equipment left behind must be destroyed by someone.
- Dead must be left behind. Someone must remove any classified documents such as signal operating instructions (authentication tables, call signs, frequencies, etc.), notebooks, and maps.

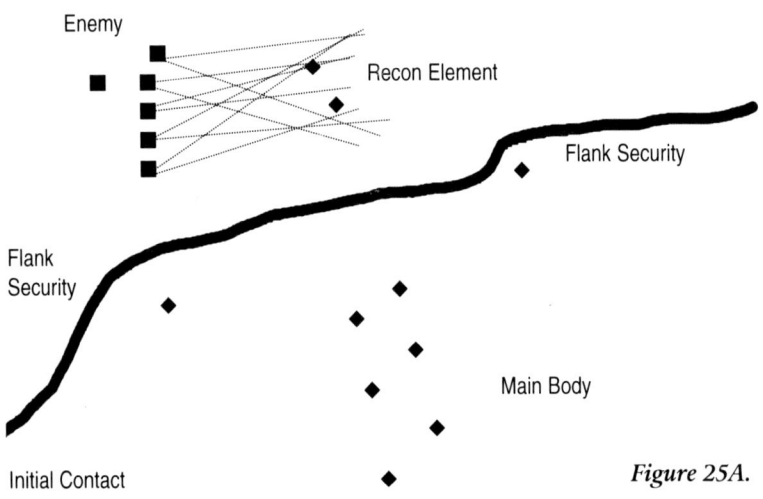

Enemy

Recon Element

Flank Security

Flank
Security

Main Body

Initial Contact

Figure 25A.

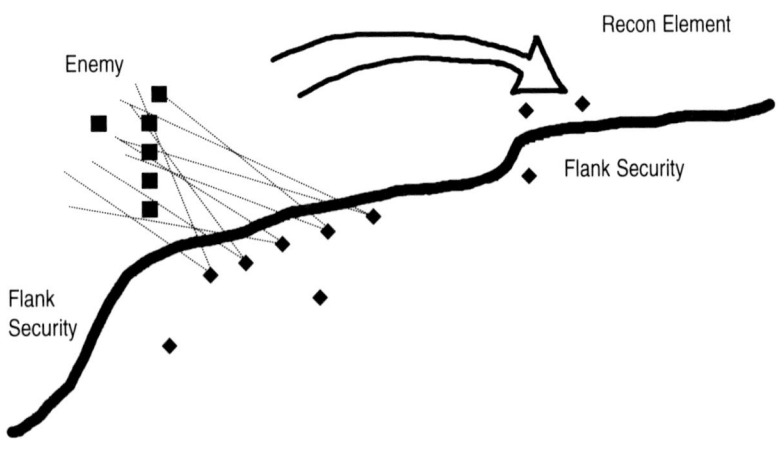

Recon Element

Enemy

Flank Security

Flank
Security

Initial Contact

Figure 25B.

108

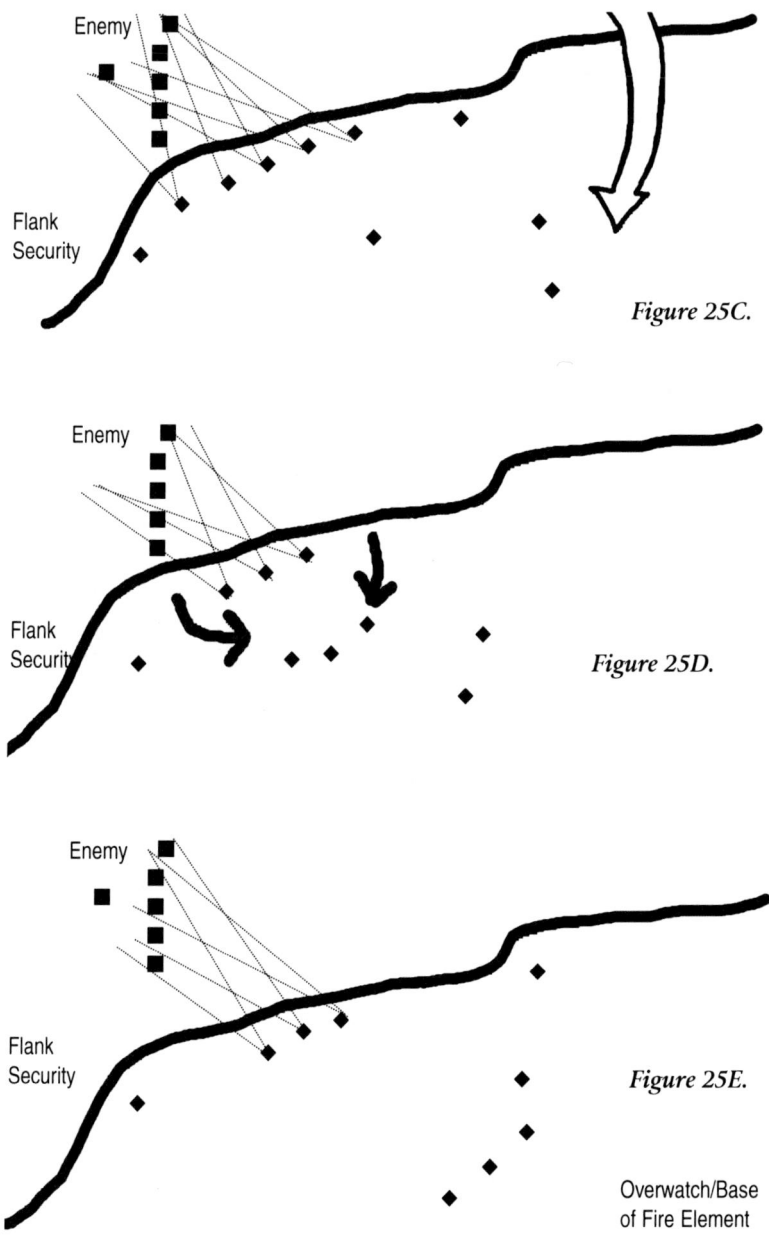

Enemy

Flank
Security

Figure 25C.

Enemy

Flank
Security

Figure 25D.

Enemy

Flank
Security

Figure 25E.

Overwatch/Base
of Fire Element

109

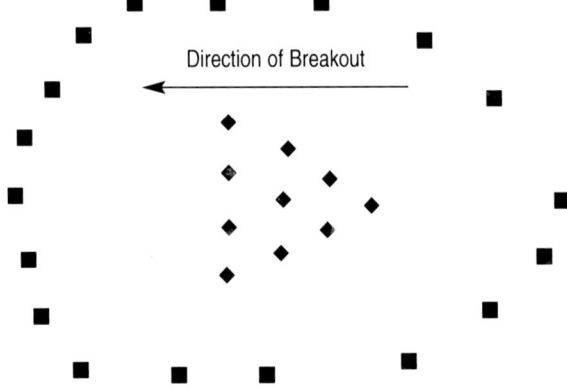

Figure 26. Patrol forms a pyramid with the base leading.

Rear Security

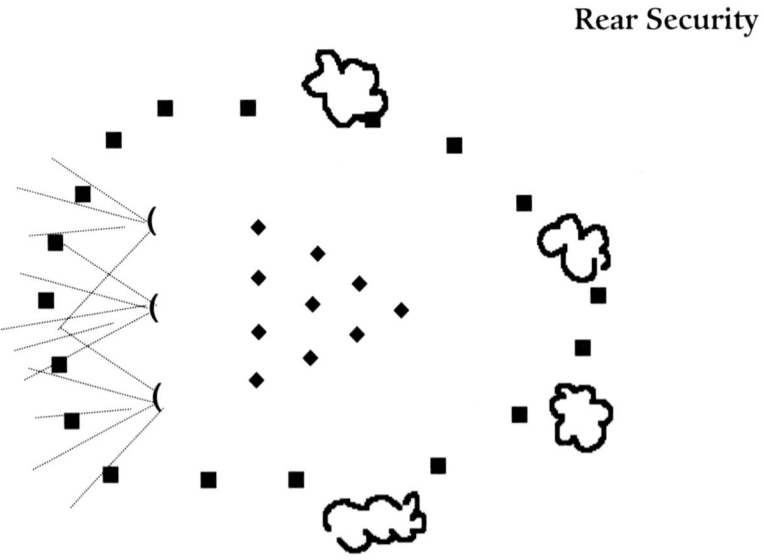

Figure 27. Grenades are thrown to rear and flanks. Claymores and grenades are discharged in the direction team will move.

110

- Depending on the size of the patrol, one or two persons must have the responsibility of rear security. This will include assisting any personnel who may be wounded before or during the attempt. Additionally, they should recover documents from members killed during the breakout movement. No attempt should be made to take KIAs with the team.

Formation for Breakout

The most effective formation for a small team (5 to 12 men) is to form a pyramid configuration, with the base leading. The following actions should take place (figs. 26-30):

- The team forms into position.
- CS rounds from grenade launchers are fired or thrown to the flanks.
- White phosphorus grenades are thrown to the rear.
- A claymore is fired or high explosive grenade is thrown in the direction the team will move.
- Immediately after the claymore/grenade detonates, the team moves out.
- The first element of the line will fire full automatic. The remainder of the team holds fire.
- When the first element's magazines are empty, the second element moves through them and continues the fire.
- When the second element has emptied its magazines, the first element will have reloaded and will pass through them and take up the assault, but firing only on semiauto.
- The team must move rapidly but should not run. It never stops until it is completely out of the encirclement.

Supporting Fires

When indirect fire support is available, the patrol leader always integrates it into his plan. When breaking out from encirclement, troops are taught to move out of the impact

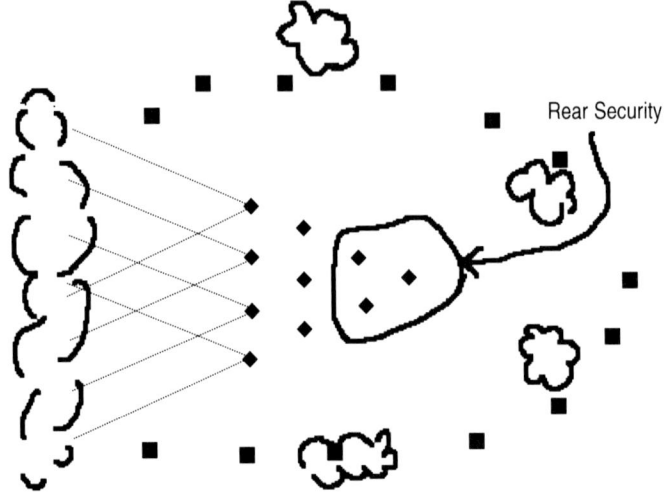

Figure 28. Immediately after the frontal explosion, the team moves out. The front rank fires on full automatic. The remainder of the team holds its fire unless a target poses a threat.

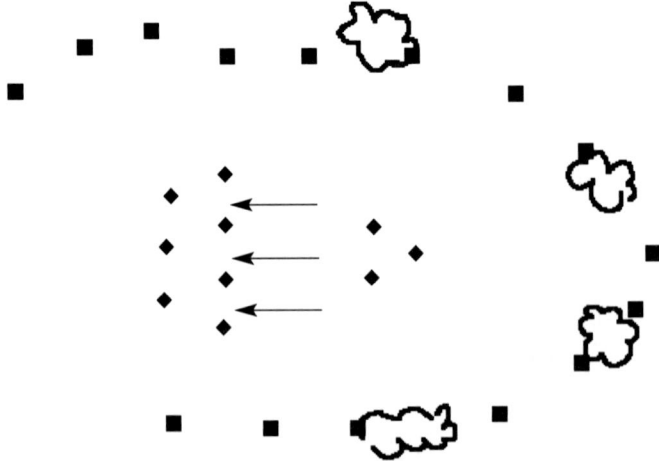

Figure 29. When the lead element's magazines are empty, the second rank moves through them and continues to fire.

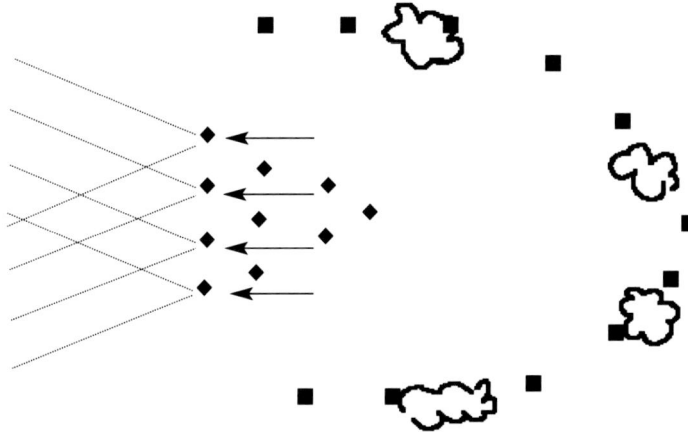

Figure 30. When the second rank has emptied its magazines, the first rank (having reloaded) passes through firing on semiauto.

area of indirect fire weapons. First, have the fires placed completely around your position. Then, having selected your desired heading, "walk" the fires in front of you. This will effectively lead you out of the danger area, and you may pick up a shell-shocked prisoner on your way out.

CHAPTER 3

PATROL BASES

A patrol base is a position established when a patrol halts for an extended period, usually not to exceed 24 hours. It is occupied the minimum time necessary in order to accomplish the purpose for which it is established. The same patrol base is not used again at a later date.

When a patrol is required to halt for an extended period in an area not protected by friendly troops, it must take measures to provide maximum security while in such a vulnerable situation. The most effective means is to select an area that provides passive security from enemy detection. In guerrilla warfare, the insurgent is usually at a disadvantage when indirect fires, air power, or ground attack is concerned. For this reason, a patrol base is evacuated if discovery is even suspected.

Planning for establishment of a patrol base is usually part of the patrol's operation order. Sometimes, however, this may be an on-the-spot decision after the recon, securing, and organization of an area during a security halt. Typical situations that require planning for a patrol base include a need to:

- Cease all movement to avoid detection.

- Hide the patrol during a lengthy, detailed recon of the objective area.
- Prepare food.
- Prepare a final plan and issue orders prior to actions at the objective.
- Conduct extended operations or several operations before returning to base.

The selection of a patrol base is usually made by map recon during the planning phase of the patrol, by aerial recon, or from prior knowledge of a suitable location. A patrol base that is preplanned prior to the patrol is tentative until it is determined to be suitable and secured.

Plans to establish a patrol base must include selection of an alternate location, a rendezvous point, and a rally point.

The alternate patrol base is used if the initial location proves to be unsuitable or if the patrol is required to evacuate the initial location prematurely. If it is feasible, the alternate location should be reconned and kept under surveillance until occupied or no longer needed.

The rendezvous point is used if the patrol evacuates the patrol base in small groups. It will not have been reconnoitered.

The rally point is used if the patrol is dispersed from the patrol base. It is a point over which the patrol has previously passed, has found suitable, and is known to all.

CHARACTERISTICS OF A PATROL BASE

Select:
- Terrain that would be considered of little tactical value.
- Difficult terrain that restricts foot movement.
- Dense vegetation that provides close ground cover and concealment from air observation (if possible).
- An area away from human habitation.
- An area located close enough to water for resupply but

116

not in the immediate area. Enemy recon patrols may watch water.

Avoid:
- Known or suspected enemy locations.
- Built-up areas.
- Ridge lines and topographic crests.
- Roads, trails, and natural lines of drift.
- Wet areas, steep slopes, and small valleys, which may be lines of drift.

Plan for:
- Outpost and listening post systems covering avenues of approach into the area.
- Communications with outposts and listening posts to provide for early warning of enemy approach.
- Defense of the patrol base if required.
- Withdrawal, to include multiple withdrawal routes.
- An alert plan that ensures a predetermined number of personnel are awake at all times.
- Enforcement of camouflage, noise, and light discipline.
- Conduct of activities with minimum movement and noise.

OCCUPATION OF A PATROL BASE

A patrol base may be occupied in either of two ways: by moving to the selected site and expanding into and organizing the area in the same manner as an on-the-spot establishment, or halting short of the selected site and sending forward reconnaissance forces. The method used must be planned and rehearsed thoroughly. Performing patrol base drills will assist in the desired stealth by allowing for swift and efficient occupation.

The following is an example of a platoon-size patrol base occupation and operation.

117

Approach
- Patrol is halted outside of sight, sound, and small arms range of the tentative base.
- Close-in security is established.
- Squad leaders accompany patrol leader to recon the site.

Reconnaissance
- Patrol leader designates a point of entry into the patrol base as 6 o'clock.
- Patrol leader moves to and designates center of base as headquarters.
- Squad leaders recon areas assigned by the clock system for suitability and return to patrol leader.
- Two members are dispatched to bring the patrol forward.

Occupation
- Patrol leaves line of march at a right angle and enters base in single file, moving to the center of the base. Designated men remove signs of patrol's movement.
- Each leader peels off his unit and leads it to the left flank of the unit's sector.
- Each unit occupies its sector by moving clockwise to left flank of next sector.
- Patrol leader checks the perimeter by meeting each leader at the left flank of his sector and moving clockwise. Sectors of fire are checked as well as placement of individuals and weapons.
- Each leader recons forward of his sector by moving a designated distance from the left flank of the sector, moving clockwise to the right limit of his sector (fig. 31). He reports indications of enemy or civilian activity, suitable observation post positions, rally points, and withdrawal routes.
- Patrol leader designates rally points, positions for observation and listening posts, and withdrawal routes.

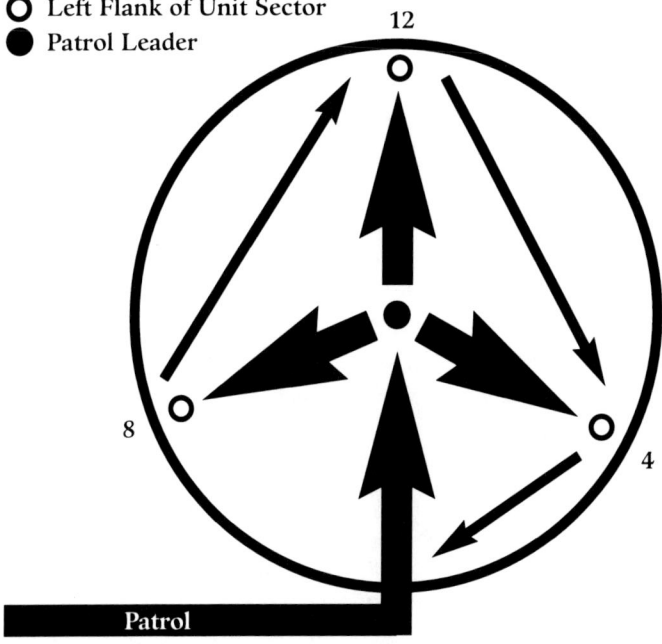

○ Left Flank of Unit Sector
● Patrol Leader

Figure 31.

- Each unit puts out one two-man observation post (day) and one three-man listening post (night) and establishes communications.

Operation
- Only one point of entry and exit is used. This point is camouflaged and guarded at all times.
- Fires are built only when necessary and, as a general rule, only in daylight. Whether day or night, fires are kept as small as possible. Where terrain permits, fires are built in pits and covered and shielded. Dry, hard wood must be used to reduce smoke. One way to dissipate smoke is to

119

dig a trench up the slope, then cover it, leaving small holes in several places to spread the smoke.

- Noisy tasks are accomplished only at designated times and never at night, early morning, or evening. When possible, use distant battle noises to mask patrol base sounds.
- Movement, both inside and outside the base, are restricted to a minimum.
- Civilians who discover the patrol base are detained until the base is moved or evacuated. Care is taken to ensure detained civilians learn nothing of plans or operation of the base. They are blindfolded, and their ears are covered.
- If possible, listening and observation posts are manned by two or three people so they can rotate shifts and keep movement to a minimum.
- A one-hour stand to (100 percent alert) is held every morning 1/2 hour before first light and every evening 1/2 hour before last light. No movement or noise is made at this time. All personnel cover their sector of fire.
- Each man ensures that he knows the position of everyone to his left, right, front, and rear and the times and routes of any planned movement within, into, or out of the base.
- In areas where the ground cover is dense, the buddy system may be used to remove individual leaves and other small obstacles from ground level up to about knee height in sectors of fire. This enables the prone patrol member to see the legs of approaching enemy while preventing the standing enemy from seeing through the cleared sector.

Defense

- Defense is planned for, but a patrol base is defended only when evacuation is not possible.
- Elaborate firing positions are not constructed.
- Camouflage and concealment are stressed.
- Artillery and mortar fires may be planned if available.

120

- Early warning devices may be placed on avenues of approach.

Communications

- Radios are an excellent form of communications but can be intercepted easily and must be controlled carefully.
- In conventional warfare, communications may be established with higher headquarters. In guerrilla warfare, the enemy's radio direction-finding capability and possible use of scanners must be considered.
- Communications of some sort must be established with the observation and listening posts. The system must alert every man quickly and quietly.
- Wire can be used within the patrol base if the bulk, weight, and time required to lay it and pick it up are not disadvantages. Bear in mind that the enemy could detect it and follow it from an observation or listening post directly to the patrol base.
- Tug or pull wires may be used for signaling. They are quiet and reduce radio traffic.
- Messengers may be used within the patrol base.

Maintenance

- Weapons and equipment are cleaned and maintained as required. No more than 1/3 of the personnel should be allowed to disassemble their weapons at one time. This means 1/3 of a given squad within a platoon. A single squad does not disassemble its weapons at the same time; rather, one out of three in the entire platoon does so. This ensures that all sections of the perimeter are covered by weapons.
- Cathole latrines are used for several reasons. They are easy to cover and camouflage upon evacuation of the patrol base; they reduce smell, thus lower the chance of detection; and since each person covers his own waist

after using the cathole, they help prevent the attraction of flies, which enhances hygiene.

- To dig a cathole, carefully remove the top layer of soil, trying not to damage the roots of any plants. Save this topsoil to cover and camouflage the hole later. Place the top soil in a location where it will not get stepped on.
- In daylight, catholes outside the perimeter are used. The user must be guarded. At night, catholes inside the perimeter are used.
- Cans and other trash are either buried or carried out. If not buried rather deep, small animals will dig up the cans and trash, which usually contain remnants of food, after humans leave. If stealth is required, carry out the trash.

Eating
- Men eat at staggered times.

Water
- If possible, watering spots are not used since the enemy often watches water; plus, enemy troops or civilians may use the water source.
- If it becomes necessary to go to water, water parties are organized. Security is established first—the water party observes the water spot for several minutes prior to sending anyone forward. Actions to be taken are planned if chance contact is made.
- No more than two visits to a water source are made in a 24 hour period.

Rest
- Rest and sleep are permitted only after all work is accomplished. Rest periods are staggered so that proper security is maintained. Consistent with work and security requirements, each man is scheduled to get as much sleep and rest as possible.

PLANNING AND CONDUCT OF OPERATIONS

- Details of operations must be made known to all men without assembling everybody at one time, thus endangering security. Rehearsals are limited to terrain models. Weapons are not test fired.

Departure

- All signs of the patrol's presence are removed or concealed. This is to deny the enemy knowledge of your presence and how you operate patrol bases, and prevent pursuit.

123

PATROL TIPS

Many small details of small unit patrolling and combat are best characterized as "miscellaneous" points. This does *not* mean that every one of them are unimportant, because they are. Every member of a guerrilla unit should be familiar with the following patrol tips.

PATROL TIPS

- Tape the muzzle of your weapon to keep out water and dirt. Leave the lower portion of flash suppressor slits open for ventilation.
- The last three rounds in every magazine should be tracer. This tells you when the magazine is almost empty.
- One of the magazines in a pouch should be taped for quick and easy removal. Magazines are placed upside down in the pouch with the rounds pointing away from you.
- Tape anything that rattles or otherwise makes noise when you jump up and down.
- Put cloth tape on stocks of weapons to deaden the sound of limbs hitting them (bow tape works good).

- Tape sling swivels or remove them from the weapon.
- The patrol leader may carry a magazine full of tracers to designate targets with.
- Quietly replace the cartridge in the chamber of your weapon every morning. Condensation could cause a malfunction.
- Carry a small container of oil and a small rag for your weapon.
- Lubricate the selector switch on your weapon daily.
- Everyone carries his weapon on safe, with the possible exception of the pointman. (This is up to the patrol leader.) Pointman may carry his weapon on full automatic in case of chance enemy contact.
- Clean your weapon daily. This may mean simply field stripping it and wiping with an oiled cloth, then running a bore brush down the barrel, then wiping the chamber.
- Always carry a knife on patrol.
- Use paper tape to secure grenade rings to the grenade body to prevent noise and secure the pin.
- On patrol, move 10 to 15 minutes, then halt for 5 minutes at irregular intervals. These are security halts. Everyone remains quiet, listens, and observes for movement.
- Never break limbs or branches because it leaves a trail for the enemy to follow.
- On a squad-sized patrol, only two or three members eat at once. All others are in a defensive posture. Don't eat while moving. Check your area for any dropped food.
- Check your pointman regularly. Change directions often. Divide your patrol's direction of movement into legs or sections that are designed to change directions. Never take a direct route to your objective.
- Do not return along the same route you took to an objective.
- In some climates or weather conditions, it may be prudent to do most of your moving early in the morning to con-

serve water. Don't be afraid to move at night, but remember that noise travels farther at night.

- When the patrol stops for more than a few minutes or so, use the fan technique to check 40 to 60 meters in all directions (hand grenade range).
- Each man in the patrol must visually check with the man to his front and rear every few seconds in addition to watching for other members' hand and arm signals.
- During dry seasons, don't urinate on rocks or leaves. Urinate in a hole or small crevice. The wet spot may be seen, and the odor will carry further.
- Each patrol member should carry a pair of extra large socks to be used over boots when crossing roads or trails.
- During rest halts, do not remove your rucksack or leave your weapon alone. During long halts, don't remove your pack until the perimeter is checked for 40 to 60 meters all around.
- Throw nothing on the ground.
- Don't set patterns, like always button-hooking to the left when setting an ambush on your own trail.
- Do not fire your weapon if the enemy is searching for you at night. Avoid the use of hand grenades at night if the enemy is searching for you

Figure 32. This figure shows the path of movement of a patrol that feels it may be being followed. The unit moves straight for awhile, then loops around to an ambush position along the route it just traveled. Do not set a pattern or always loop in the same direction. Try to choose the ambush positions as you move past them the first time.

127

unless they have a time delay of 60 to 120 seconds. This gives the patrol time to move.

- Buddy teams always stay together. Don't even use the latrine without your buddy when on patrol.
- Don't bend the pins flat on grenades. They are too hard to pull.
- Inspect grenades daily. Fuses can come unscrewed easily.
- Always have a primary and alternate rally point selected.
- Select water points when planning your patrol. Get water from places that are difficult to observe, such as in a bend of a stream.
- Always carry maps and notebooks in a waterproof container on patrol. Use pencils, not ink pins. Ink runs when wet.
- Inspect each team member's weapon and equipment before departing on a mission. Quiz all members on his knowledge of the mission.
- When changing socks, remove only one boot at a time. Only two patrol members change socks at one time. Try to wait until you stop for the day to change socks.
- Take good care of your feet. Try to keep a dry pair of socks handy. Remove wet socks, let feet dry, massage feet, then put on dry socks.
- Leather gloves will help protect your hands from thorns or bushes.
- Avoid overconfidence. It leads to carelessness.
- Maintain noise discipline at all times.
- No lights, cigarettes, or fires at night. Smoking while on patrol is never a good idea; it can be smelled over a great distance.
- Use camouflage discipline. Renew camouflage in the morning, at noon, and at the RON (remain overnight) or ambush site.
- If the enemy conducts security sweeps along roads, they will probably establish patterns as to the general time of day and the distance on each flank. Become familiar with

128

the way the enemy conducts security sweeps. If you have to cross a road near the enemy, try to wait until last light to do it.

- If the enemy is trying to locate your exact position at night, he might move as close as he can, then have someone farther back start shooting, hoping you will fire back and give away the patrol's position.

RON TIPS

- Select a tentative RON site from a map at least two hours in advance.
- After passing a suitable RON site, fish hook into an ambush position along your route of march so you can observe for trackers (fig. 33). After a suitable time, move into your RON.
- When in position, all team members should keep all equipment on and remain alert until the perimeter is cleared for 40 to 60 meters 360 degrees.
- Packs should not be removed until dark.
- When setting up a RON site, place the pointman opposite the most likely avenue of enemy approach so he can lead the patrol out in an emergency.
- Do not send radio messages from or near the RON site. If you do, move.
- Never cook in the RON.

Route of March

Figure 33.

129

- Never eat in the RON.
- Prior to dark, the patrol leader tells all members the primary and alternate rally points.
- Half the team should have their compasses set on the primary rally point. The other half sets their compass to the alternate rally point. If the enemy comes from the direction of the primary rally point, several compasses will be set on the alternate.
- The pack can be used as a pillow, but make sure the straps are up so the arms can be inserted easily for rapid withdrawal.
- If a person coughs or talks in his sleep, have him sleep with a gag on.
- Know what your following day's plans are before you settle down for the night.
- If you place claymores around your RON, they should be placed one at a time by two men. One man guards while the other places the mines. Never place a claymore in a position that prevents you from having visual contact with it.
- Claymore firing wires should not lead directly back to the team. Position the claymore such that if the enemy turns it around along the wires' route, it will not point directly at the team.
- Determine ahead of time who will fire each claymore and who will give the signal to fire.
- If the enemy is moving "on line" toward the RON site, they probably will not know its exact position. If rifles, claymores, or grenades are fired, the enemy may try to flank the patrol and box it in.
- All team members should be awake, alert, and ready to move 30 minutes prior to first light.
- Prior to moving out or retrieving claymores, a check 360 degrees for 40 to 60 meters should be made.
- A thorough search of the RON site is made to make sure nothing is left and the site is sterile.

130

- Be alert when leaving your RON site. If you have been detected, you will most likely be hit within 300 meters.
- Do not form patterns that the enemy can exploit such as always taking breaks at the same time or always moving into the RON in the same way or same time.

INTELLIGENCE

S un Tzu said, *"Raising a host of a hundred thousand men and engaging them in war entails heavy loss on the people and a drain on the resources. The daily expenditure will amount to a thousand ounces of silver. There will be commotion at home and abroad, and men will drop out exhausted.*

"Opposing forces may face each other for years, striving for the victory which may be decided in a single day. This being so, to remain in ignorance of the enemy's condition simply because one grudges the outlay of a hundred ounces of silver is the height of stupidity.

"One who acts thus is no leader of men, no present help to his cause, no master of victory. Thus, what enables the wise commander to strike and conquer, and achieve things beyond the reach of ordinary men, is foreknowledge. Now this foreknowledge cannot be elicited from spirits; it cannot be obtained inductively from experience, nor by any deductive calculation. Knowledge of the enemy's dispositions can only be obtained from other men.

Hence the use of spies, of which there are five classes:

1) *Local spies—Having local spies means employing the services of the inhabitants of an enemy territory.*
2) *Moles—Having moles means making use of officials of the enemy.*

3) *Double agents—Having double agents means getting hold of the enemy's spies and using them for our own purposes.*

4) *Doomed spies—Having doomed spies means doing certain things openly for purposes of deception and allowing our spies to know of them and report them to the enemy (in other words, spies that are considered expendable and thus are given fabricated information).*

5) *Surviving spies—Surviving spies are those who bring back news from the enemy's camp.*

When these five kinds of spy are all at work, none can discover the secret system. This is called 'divine manipulation of the threads.' It is the commander's most precious faculty. Hence it is that which none in the whole army are more intimate relations to be maintained than with spies. None should be more liberally rewarded. In no other fields should greater secrecy be preserved.

"(1) Spies cannot be usefully employed without a certain intuitive sagacity; (2) They cannot be properly managed without benevolence and straightforwardness; (3) Without subtle ingenuity of mind, one cannot make certain of the truth of their reports; (4) Be subtle! Be subtle! And use your spies for every kind of warfare; (5) If a secret piece of news is divulged by a spy before the time is ripe, he must be put to death together with the man to whom the secret was told.

"Whether the object be to crush an enemy, to storm a territory, or to kill an enemy general, it is always necessary to begin by finding out the names of the attendants, the aides-de-camp, and door-keepers and sentries of the general in command. Our spies must be commissioned to ascertain these.

"The enemy's spies who have come to spy on us must be sought out, tempted with bribes, led away, and comfortably housed. Thus they will become double agents and available for our service. It is through the information brought by the double agent that we are able to acquire and employ local and inward spies. It is owing to his information, again, that we can cause the doomed spy to carry false tidings to the enemy.

"Lastly, it is by his information that the surviving spy can be used on appointed occasions. The end and aim of spying in all its five

varieties is knowledge of the enemy; and this knowledge can only be derived, in the first instance, from the double agent. Hence it is essential that the double agent be treated with the utmost liberality.

"Hence it is only the enlightened and wise general who will use the highest intelligence of the army for purposes of spying and thereby they achieve great results. Spies are the most important asset, because on them depends an army's ability to march."

Chang Yu (Sung Dynasty), interpreting Sun Tzu, said, *"In our dynasty Chief of Staff Ts'ao once pardoned a condemned man whom he then disguised as a monk and caused to swallow a ball of wax and enter Tangut. When the false monk arrived he was imprisoned. The monk told his captors about the ball of wax and soon discharged a stool. When the ball was opened, the Tanguts read a letter transmitted by Chief of Staff Ts'ao to their Director of Strategic Planning. The chieftain of the barbarians was enraged, put his minister to death, and executed the spy monk. This is the idea. But expendable agents are not confined to only one use. Sometimes I send my agents to the enemy to make a covenant of peace and then I attack."*

WHAT IS INTELLIGENCE?

I don't think I could have said it as well as Sun Tzu did 2,000 years ago, but I can expand on it. In *any* type of warfare, intelligence about the enemy is paramount. If you are not doing a good job of it, you will lose.

Information is not intelligence—information is just something someone said. If you can confirm the information through other independent sources, it becomes intelligence. This chapter is dedicated to the methods of how to obtain and process information into intelligence.

In the military, it is not enough to simply have the proper clearance to get intelligence on a subject. You must have *the need to know*. Intelligence is provided only to those who need it in order to accomplish their mission. Intelligence matters must be held in secret because if the enemy knows you have a certain piece of information, that information becomes worthless to you. It can also compromise your sources or, even worse, cause the enemy to change his actions, which can lead to your defeat.

ORGANIZATION

Intelligence is not something deduced; it has to be sought, and it has to be worked for. This means it must be planned and controlled.

As in any operation, planning cannot be overemphasized. Intelligence operations are the responsibility of the resistance intelligence officer, or G-2.

The following functions are within the organization of the intelligence section:

Underground. This is the branch that consists of spies, informants, their handlers, and the operatives that perform such tasks as sabotage, infiltration of the enemy's institutions, mob control, and assassination.

Information Officer. The information officer is responsible for such things as psychological operations, providing information and misinformation, and spreading propaganda to the people and the outside world.

Military Intelligence. It is this group's job to gain and maintain intelligence on enemy military units, their emplacements, methods of operating, commanders, weapons, logistics, strengths, weaknesses, and what is termed the enemy's "order of battle."

Counterintelligence. Methods and procedures must be devised and enforced to prevent or minimize the enemy's ability to develop intelligence about the resistance.

Cell Organization

The underground is primarily organized into cells (fig. 48). The reason for this organization is security. The individual agent does not know the other agents; he has operational contact with the cell handler only. The cell handler manages the cell. At most, only four people can be compromised.

The cell handler reports to a network manager through a "cutout" (fig. 49). Net managers get their orders and direction from the area underground director. The director reports to the area commander.

CLANDESTINE COMMUNICATIONS

All communications between cell handlers and the net manager is through the cutouts. All communications are clandestine. None of the individuals know each others real names or identities.

Here is an example of a possible communications technique. The cell handler knows that he is to watch for a mark of a certain color on a certain day at a certain location. If that

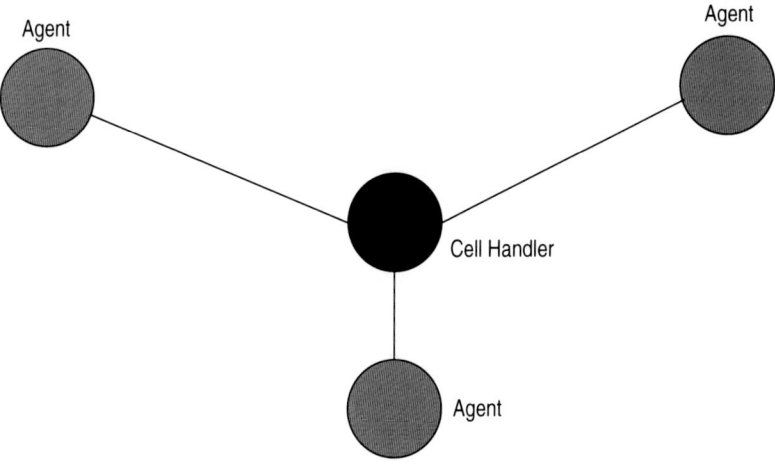

Figure 48.

137

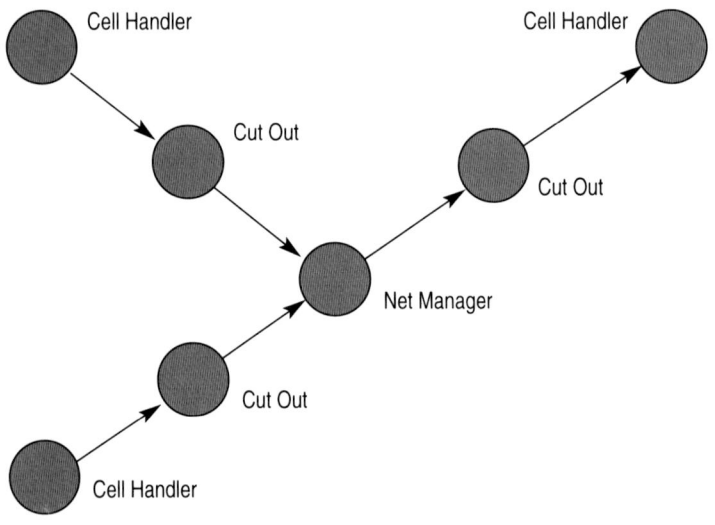

Figure 49.

mark is present, he must pick up a message at a secret location. This message will be left at a hiding place that is known
to him, such as behind a loose brick. This is known as a
"dead letter drop." This message may contain instructions or
a requirement for information. Often this message will contain instructions on the location and marking signal of the
next dead drop.

The person that drops the message may go to a distant
location to casually observe the drop site to ensure that the
message is picked up within a certain window of time and
that the handler has not been followed. If it is not picked up
at the proper time, it is considered null and void.

Sometimes cells can be given instructions or signals via a
radio broadcast. The cell member would listen for a code at a
certain time on a known frequency.

Often, it may be necessary to pass items between the cell

138

and the net manager. Instructions may be given at a drop to meet someone at a certain place at a specific time. Instructions would include "all clear" and authentication codes. The information is passed between the operatives in a way that raises no suspicion. Usually, the person that takes on the function of cutout in this situation is not the usual cutout but a courier whom the cell handler has never seen.

To illustrate this, the following example is submitted. The cell handler has become aware of enemy plans that will directly affect resistance operations. The standard operating instructions for the network provides for priority communications between the handler and net manager by a prearranged signal. This is a clandestine signal that is monitored perhaps daily to tell either party that a priority communication is required. In order to keep it secret, this means is rarely used.

A message is passed from the handler to a courier acting as a cutout after observing proper authentication signals and codes. The exchange may be monitored by the net manager from a distance to ensure that the transfer is not compromised in any obvious way.

If the net manager feels that the transfer was not compromised, he leaves a signal to indicate to the courier that he can transfer the message. If the signal is not present, the courier goes to an alternate signal location at a designated time to look for the signal.

Upon recognizing the all-clear signal, the courier leaves a signal at another location to indicate that he feels that he has not been compromised and has recognized the all clear left by the net manager. The courier then performs the transfer to the net manager using a preplanned technique such as dead letter drop or face-to-face exchange using proper recognition codes and authentication phrases.

From the time this operation is started, all personnel use evasive techniques to determine if they are being followed and to prevent it. Individuals act casual and do not take actions that would raise suspicions, even if they are being watched.

Any signal left should be made in a preplanned way that would be difficult to recognize if someone were watching. For example, while palming a small piece of specific colored crayon, the person leaving the mark stops to pull up his sock. While doing so, he rests his hand against the wall to balance himself, leaving a small colored mark as the predesignated signal. Similarly, when an individual looks to see if a signal has been left, it should not be obvious.

This procedure takes time, because this type of operation should not be hurried. If a member of the underground is compromised, he can be captured at best. At worst, he will not know of the compromise, and others could be compromised. This can make the cell ineffective for an extended period.

ESTABLISHMENT OF INTELLIGENCE NETWORKS

The resistance intelligence officer divides the operational area into logical zones. These zones will probably be along the lines established by the resistance command.

The intelligence officer places a trained intelligence operative in charge of the underground in each zone with the mission of establishing and maintaining an effective intelligence-gathering organization in that zone. We will call this operative a "zone leader."

The zone leader establishes his staff and develops a detailed written assessment of known information about the area, to include, but not limited to:

- Enemy leaders
- Military installations and units (to include unit designations)
- Potential targets

- Climate
- Industry
- Transportation
- Political, economic, and social problems and strengths
- Enemy strengths and weaknesses
- Educational institutions
- Religion and religious leaders
- Ethnic makeup
- Local attitudes toward the enemy
- Communication facilities
- Geography
- History
- Local leaders

All information is categorized and indexed in a form that will allow quick and easy retrieval and update. Primitive conditions may dictate keeping records in primitive ways such as notebooks or card files, although the use of computers in some areas may be possible. Whatever methods are used, they must be carefully controlled, and there should be a backup. They should also be capable of being destroyed easily and quickly.

Information about enemy military units should be plotted on a map using standardized military symbols. This information should be maintained and kept as up to date as possible.

Zones can be logically divided into smaller areas. These areas can be given names. These names are used to store information about the areas for easy retrieval. As an example, if the information is to be stored in an indexed file, it is stored by the given name for easy access.

Recruitment

In each area, spotters will be placed. Spotters live in the area and either are or become intimately aware of the people in the area who may be inclined to be sympathetic to the resistance. They also look for individuals who are aggressive in their opposition to the resistance. The people the spotter looks for are potential agents in the network or potential targets for termination. Potential agents must be intelligent and motivated ideologically against the enemy.

The spotters also look for people who are in influential or informed positions who could be bribed or coerced into cooperating. Spotters take note of vulnerabilities of enemy leaders.

Tu Yu said, *"We select men who are clever, talented, wise, and able to gain access to those of the enemy who are intimate with sovereign and members of the nobility. Thus they are able to observe the enemy's movements and learn of his doings and his plans. Having learned the true state of affairs, they return to tell us. Therefore they are called 'living' agents."*

Tu Mu said, *"These are people who can come and go and communicate reports. As living spies we must recruit men who are intelligent but appear to be stupid; who seem to be dull but are strong in heart; men who are agile, vigorous, hardy, and brave; well-versed in lowly matters and able to endure hunger, cold, filth, and humiliation.*

"Of all those in the army close to the commander, none is more intimate than the secret agent; of all rewards none are more liberal than those given to secret agents; of all matters, none is more confidential than those relating to secret operations."

Sun Tzu said, *"If plans relating to secret operations are prematurely divulged, the agent and all those whom he spoke of them shall be put to death."*

The spotter must not let it be known that he is a spotter. He must not raise any suspicion.

The spotter only identifies candidates. He gives information about the candidates to a recruiter. Such candidates can be classed as sympathetic and unsympathetic.

Sympathetic

Sympathetic candidates are those who are not aligned with the enemy and tend to dislike the enemy. They include:

142

- Persons who have a relative or friend who has been harmed by the enemy.
- Persons who have seen injustice and mistreatment by the enemy.
- Loners who have few friends but tend to sympathize with the resistance or are disaffected by the enemy government.
- Prostitutes who are patronized by enemy leaders or soldiers.
- Intellectuals who realize the harm the enemy oppression is causing to the people.
- Government officials who are worthy men but may have been deprived of office. Others who have committed errors and have been punished. Greedy officials who have remained too long in a lowly office. Those who have not obtained responsible positions, and those whose sole desire is to take advantage of times of trouble to extend the scope of their own powers. Those who are two-faced, changeable, and deceitful who are always sitting on the fence.

In the case of sympathetic candidates, the recruiter for that area attempts to recruit the individual to serve the resistance underground. The recruiter usually will try to establish a friendship with the candidate and gain his/her trust. (Recruiters should be good at dealing with people and persuasive.) From the beginning, the recruiter attempts to analyze the candidate's psychology, political inclinations, goals, fears, and ambitions. He must be careful not to patronize or be overly aggressive in developing a relationship with the candidate.

If and when the recruiter feels the candidate would be sympathetic to helping the resistance, he attempts to recruit him or her. Often the recruiter can accomplish this without the candidate realizing that it is deliberate. This is desirable for more than one reason. For security reasons, the recruiter will want to leave the area after all recruiting in that area has been accomplished (ideally, all recruitment is culminated at the same time). Also, the candidate could resent an obvious recruitment if he/she realizes this is why the recruiter was interested in befriending the candidate.

The recruiter introduces an agreeable candidate to a handler. The handler adds the new agent to his cell. The handler trains the recruits and guides them in the gathering of information, clandestine communications, countersurveillance, etc.

Unsympathetic

Unsympathetic candidates tend to be neutral or aligned with the enemy but are vulnerable perhaps due to something they have done or could be tempted into doing (financial problems, alcoholism, drug addiction, infidelity, etc.).

- Enemy leaders who can be compromised, blackmailed, bribed, or threatened.
- Enemy military personnel with weaknesses that the spotter has identified.
- Government employees in key positions such as postal personnel, clerks for leaders, couriers, security personnel, logistical personnel, communications experts, etc.
- Enemy agents. "When the enemy sends spies to pry into my accomplishments or lack of them, I bribe them lavishly, turn them around, and make them my agents." (Li Ch'uan, c. 618-905 A.D)

Unsympathetic candidates can be spotted and recruited in much the same way as sympathetic candidates, but it normally requires a different motivation, whether it be, as examples, fear of being compromised after being photographed with the wrong person or doing the wrong thing. Another approach is for the recruiter to ask a seemingly small favor that is a minor breach of security and then pay for it. The payment could be

144

much more than the information is worth, and such small favors can continue until the recruiter asks for a larger favor. If the individual refuses, he is threatened with exposure. Photographs, canceled checks, and other forms of proof can be used to put teeth in the threat. If this person does not cooperate or if the recruiter feels the individual may go to the authorities, he should be lured to a place that allows his termination and the escape of the recruiter. The recruiter will leave the area because of his association with the individual.

If the candidate has intimate knowledge of an enemy installation that is a planned target, he could give up much information to a recruiter in casual conversation. If the target is to be attacked, the candidate can be kidnapped after being lured to an area where there are no witnesses and interrogated for detailed information that will assist in the attack.

Last but not least, information can often be bought. If information is paid for but turns out to be false, action should be taken to ensure the individual understands his mistake and doesn't make it again, and that others gain enlightenment from his mistake.

Information Flow

The flow of information is from the cells up to the area command. Only the area command maintains the information and processes it into intelligence. This is not to say that at lower levels within the command individual leaders do not try to catalog information and process it mentally; they just don't process it before passing it up. The area command has much better resources to check the information for authenticity and accuracy. It can also be compared with intelligence gained from other areas to form a bigger picture of enemy strategy, capabilities, strengths, and weaknesses. Intelligence ultimately is disseminated on a need-to-know basis only to allow secure planning of operations and enhance the security of subordinate units.

Military Intelligence

Military intelligence requires a separate group to gain and maintain intelligence on enemy military units, their emplacements, methods of operating, commanders, weapons, logistics, strengths, weaknesses, discipline, and tactics. To gain information, the military branch of the resistance conducts operations. Very often, information is obtained during both offensive and defensive tactical operations.

Operations conducted with the specific purpose of gaining information consist of, but are not limited to reconnaissance and prisoner snatches.

Reconnaissance

Small teams are sent out to observe and not be detected. If they are detected, the information may not be as useful, or they may be overwhelmed by a superior enemy force. They may be sent to confirm or deny other sources of information or to gather information that will be used for the planning of local operations. Recon teams can be used to watch trails, roads, rivers, and other transportation corridors. They can watch small towns, villages, and enemy facilities and encampments. Teams make detailed notes and drawings. They note when and where things are seen or happen.

Prisoner Snatch

Teams can be sent to capture an enemy soldier or government official. Because of the need for surprise, a small team is usually used. Larger units may be stationed to support the snatch team after seizure of the target.

Usually, silent capture is desirable because of the likelihood of enemy reaction. Techniques used for the snatch will depend upon the situation, but regardless of what techniques are used, they are characterized by detailed planning and preparation, surprise, stealth, speed, overwhelming superiority, and support.

146

NOTE: One thing to bear in mind when deciding the method of disabling the target is that excessive blood loss causes shock. Shock causes death. Dead men cannot talk.

The snatch team may recon the area to determine the best location and time to execute the operation. Detailed information is needed on such things as routes and methods of enemy travel, size of patrols, size of point elements, security measures, weapons, readiness (do they carry their weapons at the ready?), alertness, discipline, and techniques for such things as gathering water (do they come alone to get water?), and relieving themselves (do they use the buddy system?). Specific questions to be answered include when does the enemy sweep roads in the area? How far do they sweep on each side? Are routes to and from enemy watering points and latrines guarded? What areas along enemy routes of travel are difficult to observe? What are the best approaches to and from the enemy routes of travel?

Once the snatch team has the information it needs, the leader finalizes plans. The team will have rehearsed the technique to be used before entering the area. The best routes of approach and withdrawal are selected, and whatever support is available is briefed.

In Vietnam, American forces had the advantage of air support. When snatch operations were conducted, we could call for extraction. This will most likely not be available to the insurgent. Instead, the enemy may have aircraft as well as reinforcements. This situation may make delaying techniques appealing. Trip wires, antipersonnel mines, indirect fire support, snipers, ambushes, and diversions are all methods of delaying enemy reaction to fire.

Some U.S. Army Special Forces snatch operations in Vietnam (and surrounding countries) used the ambush in conjunction with pure bravado to take prisoners. This was done after an extensive recon.

147

One such technique had the team leader along with another man positioned a few feet up the trail from the rest of the team, which would be laying in ambush. When a small enemy patrol walked by, the two men would move out onto the trail behind the last man. The team leader would have a baseball bat; the other man would have his weapon ready to shoot anyone who looked back. It was timed such that when the enemy patrol moved into the kill zone, the man with the bat hit the last man across the back of the shoulders very hard. He would then fall to the ground on top of the target while the ambush was sprung. All other enemy personnel were killed by the ambush and the man next to the team leader. Usually someone had a tranquilizer injection ready to sedate the target to help control him and prevent shock.

The operation was carefully planned and timed. Helicopter extraction was done as soon as possible. An observation plane and possibly helicopter gun ships and jets were waiting to provide air support. Preplanned landing zones were used (primary and alternates).

As said before, a guerrilla force is not likely to have extensive support, so good intelligence, recon, planning, rehearsals, local support, and delaying techniques take on added importance.

Prisoner Interrogation

In spite of popular opinion, interrogating an enemy soldier is not done with a big knife pressed against his throat. It is best done by a trained person or team of interrogators. With rare exceptions, a terrified person will tell you anything he thinks you want to hear to save his life. Instead, taking the person away from the immediate combat area, isolating him from people and creature comforts, providing minimal water and food, and preventing him from relieving himself or sleeping will usually weaken his will.

148

When the prisoner is ready for interrogation, one method used is the Mutt and Jeff technique. It works like this. A two-man interrogation technique is used, and their performance must be convincing. One interrogator is openly hostile and does most of the talking while the other one observes. If the prisoner does not become cooperative, any information that may be already known is used against him. He is asked questions for which the answers are known by the interrogators. Whenever he lies he is punished (*nonlethal*). He is told that they already know the answers to most of the questions, but they need him to verify certain things.

If the prisoner remains uncooperative, the aggressive interrogator pretends to make an attempt to harm him but is stopped by the silent one. The less aggressive interrogator convinces the aggressive one to leave and let him conduct the interrogation.

The interrogator tells the prisoner that he saved him this time, but he may not be there next time. In order to prevent him from getting hurt, he needs some cooperation. He tells him to give him something, no matter how small. The interrogator then asks questions that he already knows the answers to (if possible). When he starts getting truthful cooperation, he begins asking real questions. The interrogator may give small rewards for cooperation. He attempts to develop a sense of trust but maintains strict control.

After a prisoner begins to cooperate, the interrogator attempts to reconstruct as much of the detail prior to his capture as possible. One very good method of doing this is map tracking. The interrogator uses a map to locate where the individual was captured and goes backward in time, asking the individual detailed questions about every aspect of his activities. The interrogator can trace the prisoner's movements prior to capture by asking questions about key terrain features, water sources, towns, etc. The interrogator goes back as far as he can. Initially, he is interested in information

149

of an immediate nature. Later, after such information has been sent forward to be evaluated, the interrogator seeks longer term information such as the prisoner's induction into the military, type and location of training, the units he has been with and when, names of commanders and fellow soldiers, and discipline of units.

Feints and Ruses

Operations can be conducted to see how the enemy will react in order to gain information about his tactics, weapons, discipline, etc. Resistance forces can fake activities to fool the enemy and gain intelligence. Rumors can be spread by the underground and auxiliary to cause the enemy to respond. Do they act on rumors? Do they send recon teams? If the resistance does it enough, do they stop responding? When the enemy stops responding, it may signal frustration or a realization that they should react to valid intelligence and not rumors. When they realize this, there may be an extended lag in time until they can develop and process intelligence. This window may provide opportunities.

Tactical Operations

Much of the military information about the enemy is gathered during the conduct of normal combat operations. Resistance forces must be taught how to gather information about the enemy and report it to their leaders. Leaders send the information forward as soon as possible. Often, intelligence officers accompany the resistance forces on combat operations in order to conduct training, observe procedures for gathering information, and advise leaders about gathering information.

Each unit must adhere to standard procedures for gathering and forwarding information. Normally, the military intelligence group establishes these procedures and may perform audits on units (with permission of the unit commander).

Debriefing

After *all* operations, the unit goes through a debriefing. This debriefing should be conducted by a trained intelligence officer. Often, intelligence personnel are permanently assigned to units.

Debriefings are done to gather as much information about the enemy and his activities as possible. Information about the terrain, civilians, transportation routes, weather, and anything else the intelligence officer feels is valid is covered in a thorough debriefing. Photographs, drawings, terrain models, and maps are useful.

COUNTERINTELLIGENCE

Counterintelligence is comprised of actions taken to prevent or reduce the enemy's ability to gain information about the resistance. The following are miscellaneous counterintelligence issues applicable to the guerrilla warfare environment.

Need to Know. As mentioned before, even if a member of the resistance is cleared to receive a level of classified information, he is not allowed access to it unless he has the need to know.

Restricted Access. Restricting access is reducing or eliminating nonessential personnel access to or through an area. This is to prevent them from gaining knowledge of friendly actions, equipment, or installations that could later be compromised to the enemy.

As an example of this, if the resistance forces use a bordering country as a sanctuary, they should attempt to control areas of the border or make them no man's land. The movement of all personnel other than active partisan fighters in this area is made to be very dangerous. One of the reasons for this is to deny the enemy information about movements along the border. Another reason is to help prevent the enemy from isolating the resistance from their sanctuaries or interrupting the movement of supplies, wounded, or reinforcements.

Another example of restricted access is securing an area where the resistance is making some type of tactical preparation, such as an isolation area. Isolation areas are where units are taken to separate them from all others before giving them a mission to prepare for. Until they return from the mission, contact with anyone outside of the unit is forbidden to prevent security leaks.

Diaries. Soldiers are not allowed to maintain diaries. If these fall into enemy hands, they can provide valuable information to the enemy.

Letters. Soldiers should be allowed to correspond with loved ones, but it should be limited in frequency and time. If preparations are being made for an operation, they should not be allowed to write or mail letters. During times of limited activity, they may give the letters to a designated intelligence officer. This person may censor these letters to prevent sensitive information from being compromised inadvertently.

Conversation. Troops are instructed not to discuss military matters with others. If a civilian or anyone else that does not have the need to know asks questions about his unit's mission, encampments, weapons strength, etc., it should be reported.

Telephone Security. As with radios, persons that use the telephone should know that the enemy is probably listening. In some situations, computers with high-speed modems can be used. If available, some sort of good encryption software should be used.

Other Forms of Communications. When time and circumstances permit, communications other than radio should be used. Such methods include messengers, signal mirrors, flags, smoke, and homing pigeons.

If messages are very simple, they may be memorized for security, but the message is subject to distortion, and it cannot be coded very well. Most situations will call for written, coded messages.

Equipment. Troops are instructed to avoid abandoning equipment because the enemy can often gain intelligence from it or use it against them. If equipment cannot be taken with them, they destroy it to prevent its use by the enemy.

Personnel Killed in Action. Every attempt is made to not leave dead personnel behind, because in addition to the morale problems it can cause, the enemy may get valuable information from them such as physical conditioning, health, nutrition, armaments, and discipline. If he is identified, retaliation against his family or village could result. If the dead must be left, an attempt is made to sterilize them by removing anything that could provide information to the enemy.

Documentation. All documentation is strictly controlled, and if the unit is about to be overrun, it is destroyed. The preferred method of destruction is to burn it and spread the ashes.

False Information. Incorrect information can be left where it is likely to be found by the enemy. This could give false information about planned operations or personnel, or it can implicate an enemy official.

Refugee Camps. In many situations, war creates refugees. Refugees often end up in camps. These camps should be avoided by the guerrillas as a whole since the enemy can be expected to have agents in them or at least watch them. If family members or friends are in these camps, guerrillas will try to contact them. They should be restricted from these camps. Any communications with individuals in the camps should be done through members of the resistance assigned to this task. The refugees can be expected to repeat what they hear.

Local Security. Tactical units must actively patrol and observe for their own security. They must never let their guard down when in hostile territory. Resistance units must never rely on civilians to warn them of enemy approach.

153

They must have multiple avenues of withdrawal, rehearsed withdrawal plans, and a defensive plan to be used if surrounded and breakout is not immediately feasible.

Sterilizing the Area. Resistance units sterilize areas when they prepare to leave in order to make it look like they were never there. Even if the enemy does find the location, it will make discerning information from it much more difficult.

Radio Communications Security.

Radio provides a form of communications in military operations that has advantages other forms do not. But there is a price paid for this speed—security. To limit the effects of this security issue yet still use the speed of radio, the radio operator must be aware of how it becomes a security problem and take defensive measures to minimize the risk.

Early in an insurgency, the use of radios should be highly restricted, especially if the enemy has any type of sophistication or has allies that do. This is because by using radio direction finding (RDF) equipment, the transmitter can be located within a few meters within seconds under excellent conditions. There should *never* be a radio transmission from a guerrilla base unless enemy contact has been made and they already are aware of the exact location of the base.

Radio transmitters emit energy from the antenna. Depending on the design, the radio can emit energy if it is just turned on. The most significant amount of energy is, of course, emitted during transmission.

The type of antenna used helps determine the direction or directions of the bulk of the transmitted energy. Some antennas are directional (fig. 50), some are bi-directional (fig. 51), and some are omnidirectional (360 degrees, fig. 52). By selecting the correct type of antenna, most of the energy can be directed toward the intended reception station and not in other directions.

The frequency that is used makes a big difference as to the angle of energy transmitted. Higher frequency energy leaves the antenna at a flatter angle.

High frequency, or HF, radio waves have what is called a "sky wave" and "ground wave." The sky wave leaves the antenna and is refracted off the ionosphere and returns to

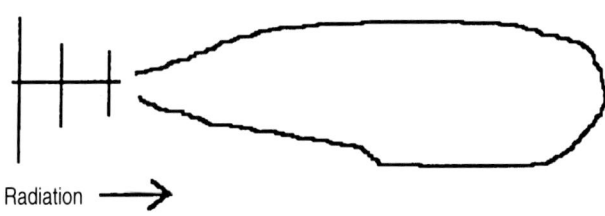

Radiation $\longrightarrow$

Figure 50. A directional antenna pattern.

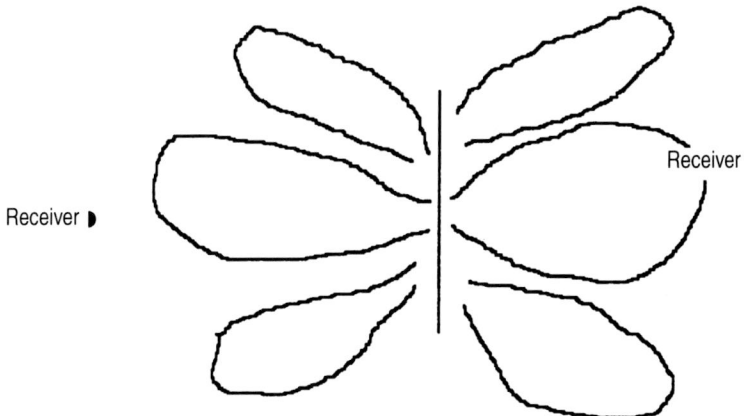

Receiver

Receiver

Figure 51. A bi-directional antenna pattern.

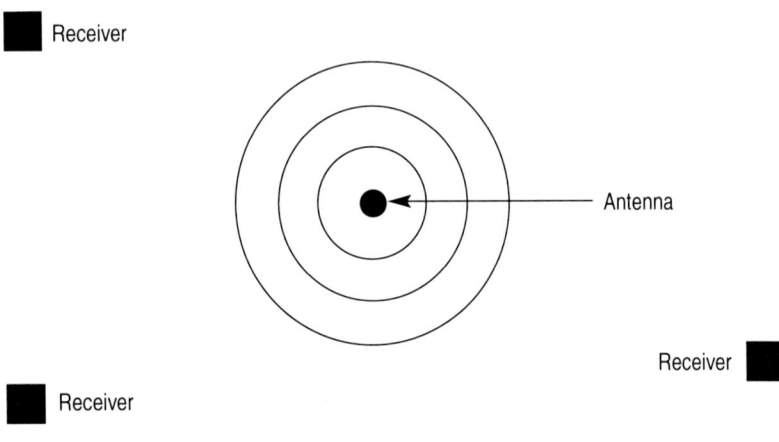

Figure 52. An omnidirectional antenna pattern (seen from above).

earth, allowing long-range communications. The ground wave leaves the antenna and travels at an angle toward the ground (fig. 53).

As shown in Figure 54, VHF frequencies transmit the bulk of their energy at a slight angle toward the ground, giving what is termed "line of sight" communications. This means that if a large mass (such as a mountain) is in the way, reception is difficult if not impossible. In the figure, communications with someone at point A would no be good. Communications with points B and C would be good depending on the distance and power of the transmitter.

The characteristics of the radiation patterns can be used to the advantage of the operator to help prevent the enemy from receiving the signal. For instance, if VHF frequencies are used, the transmitting operator can use hills and mountains to mask his transmissions.

The radio operator should *always* assume that the enemy is listening. He should keep transmissions as short as possible, not transmit from or near a base, use masking tech-

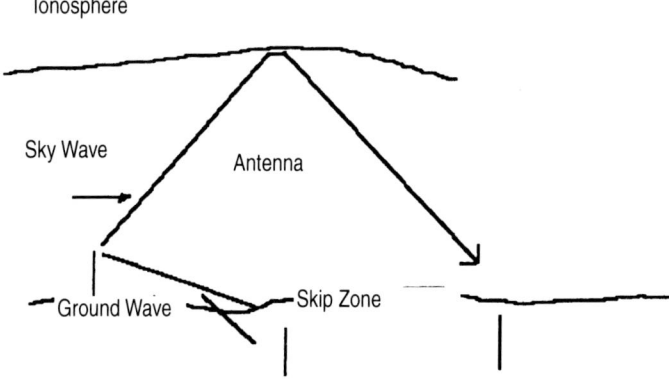

Figure 53. Sky wave and ground wave.

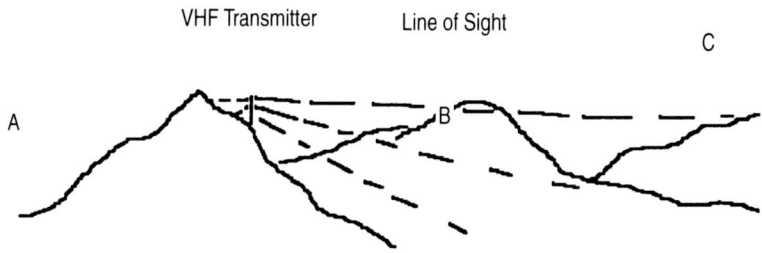

Figure 54. Line of sight transmissions.

niques, and, if possible, use directional antennas. To help keep the transmissions short and add to security, the operator should use call signs, code words, and brevity codes.

157

Call signs are meaningless names given to the unit or individuals that are known only to those with the need to know.

Code words are used to give uncommon names to things so that others will not know what is being referred to. Do not use obvious code words such as "sticks" to mean rifles or "big guy" to indicate a leader. Code words should be changed often.

Brevity codes are used most often when ordering supplies. Numbers given to items in a supply catalog are used as brevity codes. As an example, 112 could mean rifles, 111 could mean rice, 443 could indicate grenades, etc. Often a number will indicate not only the type of supplies requested but also the amount. So 167 could indicate 500 rounds of ammunition, 456 could indicate 2,000 rounds of ammunition, and 432 could mean 10 kg of rice.

If the guerrillas have the technology, burst transmissions can be used to keep the transmissions short. Packet radio is a technology that is cheap and available in many places. Packet radio uses a personal computer, special software, packet modem, and radio to send messages at much higher speed than they could be spoken.

This text does not intend to teach you the many details of how to be a radio operator. Rather, it gives an insurgent a basic understanding of the security matters involved and advises that radio operations should be done by trained personnel.

CHAPTER 6

PSYCHOLOGICAL OPERATIONS

Guerrilla warfare is essentially a political war. Therefore, its area of operations exceeds the territorial limits of conventional warfare and penetrates the political entity itself—the "political animal" that Aristotle defined.

In a political war, human beings should be considered the priority objective. Once his mind has been reached, the "political animal" has been defeated without having to resort to bullets.

Guerrilla warfare is born and grows in this environment, in the constant struggle to dominate that area of political mentality that is inherent in all human beings. It is precisely where victory or failure is defined.

This conception of guerrilla warfare as political war turns psychological operations into a decisive tool. The target, then, is the minds of *all* the population: the resistance troops, the enemy troops, and the civilian population.

In guerrilla warfare, more than in any other type of military effort, psychological activities should be conducted simultaneous with military actions in order to achieve the desired objective.

GENERALITIES

The nature of the environment of guerrilla warfare does not usually permit sophisticated psychological operations from within the area of operations. It therefore becomes necessary for the chiefs of groups and detachments to have the ability to carry out, with minimal instructions from higher levels, psychological operations.

When conducting psychological operations, one may be tempted to go to any length to put a bad face on the enemy and justify it as acceptable due to the importance of the cause. But distancing one's self from the truth invites destruction of credibility. Once credibility has been compromised, it is gone. On the other side of the coin, corrupt or tyrannical governments invite the wrath of the truth when it is presented to the people properly.

Combatant-Propagandist Guerrillas

In order to obtain the maximum results the psychological operations in guerrilla warfare, every combatant should be as highly motivated to carry out propaganda face-to-face as he is to fight in combat. This means that the guerrilla's individual political awareness of the reason for the struggle will be as acute as his ability to fight. This knowledge will not only make him an effective interface with the people but will also sustain him when he is hungry, afraid, or suffering the many discomforts associated with guerrilla warfare.

Such political awareness and motivation is obtained through the dynamics of group discussion as a standard method of instruction for the guerrilla. Group discussions raise the spirit and improve the unity of thought of the guerrilla, and they put social pressure on the weak members to perform better in future training or combat.

The desired result is a guerrilla who can persuasively justify his actions when he comes into contact with any member of the population and especially with himself and his fellow guerrillas when dealing with the challenges of guerrilla warfare. This means that every guerrilla should be able to give 5 or 10 logical reasons why, for example, a person should give him cloth, needle, and thread to mend his clothes. When the guerrilla can do so, government propaganda will never succeed in making him an enemy in the eyes of the people. It also means that, due to this constant orientation, hunger, cold, fatigue, and insecurity will have a meaning, psychologically, in the cause of the struggle.

Armed Propaganda

Armed propaganda includes every positive act carried out by a guerrilla force. The good impression that the armed guerrilla force leaves will result in positive attitudes toward that force by the population. It does *not* include forced indoctrination.

This means that an armed guerrilla unit in a rural town will not give the impression that weapons are their strength *over* the people but rather that they represent the strength of the people *against* the repressive government. This is achieved through a close identification with the people, including hanging up weapons and working with them on their crops, construction projects, and fishing; providing explanations to young men about weapons (e.g. letting them handle an unloaded weapon, describing in a rudimentary manner its operation, etc.); describing with simple slogans how weapons will serve the people to win their freedom; and working toward the people's requests for hospitals, education, reduced taxes, a just judicial system, etc.

The goal is to have the people identify with the weapons and the guerrillas who carry them so that they feel that the weapons are, indirectly, *their* weapons to protect them and help them in the struggle against an oppressive regime. Implicit terror always accompanies weapons, since the peo-

161

ple are subconsciously aware that they can be used against them, but as long as explicit coercion is avoided, positive attitudes can be achieved with the presence of armed guerrillas within the population.

Armed Propaganda Teams

Armed propaganda teams are formed through a careful selection of persuasive and highly motivated guerrillas who move about within the population, encouraging the people to support the guerrillas and resist the enemy. It combines a high degree of political awareness and the "armed" ability of the guerrillas to manage a planned, programmed, and controlled effort.

The careful selection of the staff, based on their persuasiveness in informal discussions and their ability in combat, is more important than their degree of education. The tactics of the armed propaganda teams are carried out covertly and should be parallel to the tactical effort in guerrilla warfare.

Development and Control of Front Organizations

The development and control of "front" or facade organizations is carried out through subjective internal control at group meetings by infiltrated cadres.

Established citizens—including doctors, lawyers, businessmen, and teachers—are recruited initially as "social crusaders" in seemingly "harmless" movements in the area of operations. This provides a pool of like-minded persons from which can be recruited inside cadres. When their involvement with the clandestine organization is revealed to them, it supplies the psychological pressure to continue to act as inside cadres in groups to which they already belong or of which they can be members.

After successful recruitment, they receive instruction in the techniques used to gradually and skillfully persuade target groups to support the revolution. A cell control system isolates individuals from one another.

Control of Meetings and Mass Assemblies

The control of mass meetings in support of guerrilla warfare is carried out internally through a covert group consisting of bodyguards, messengers, shock forces (initiators of incidents), placard carriers (also used for making signals), and shouters of slogans. Everything is under the control of the psychological operations command element of the resistance.

After cadres are placed or recruited in organizations such as labor unions, student groups, agrarian organizations, or professional associations, they begin to manipulate the objectives of the groups. Inside cadres help foster a mental attitude in the group that, at the crucial moment, can be turned into a fury of justified violence.

Cadres also have the mission of giving the impression that there are many of them and that they have a large popular backing. Using these tactics and a force of 200 to 300 agitators, a demonstration can be created in which up to 20,000 persons take part.

Support of Contacts in Real (as Opposed to Front) Organizations

It is important to recruit local contacts who are members of "real" organizations such as medical associations, small business groups, league of voters groups, social clubs, and other professional organizations.

COMBATANT-PROPAGANDIST GUERRILLA

An important goal of guerrilla psychological operations is to convert the individual guerrilla into a propagandist in addition to being a combatant, which maximizes the affect of a guerrilla movement. The nature of the guerrilla warfare

163

environment does not permit sophisticated facilities for psychological operations, so the effective face-to-face persuasive efforts of each guerrilla should be utilized.

Political Awareness

The individual political awareness of the guerrilla—the reason for his struggle—is as important as his ability to fight. This political awareness is achieved by:

- Developing in each guerrilla the ability to win the support of the population, which is essential for success in guerrilla warfare.
- Improving the combat potential of the guerrilla by improving his motivation for fighting.
- Teaching the guerrilla to recognize himself as a vital tie between the resistance effort and the people, whose support is essential for the sustenance of both.
- Promoting the value of participation by the guerrillas and the people in the civic affairs of the insurrection.
- Fostering the support of the population, which provides a popular psychological groundwork for politics after the victory has been achieved.
- Developing trust in the government in exile (if present) and in the reconstruction of a local and national government.

Group Dynamics

Political awareness building and motivation are attained through group dynamics at the small unit level. The group discussion method is the general technique.

Group discussions raise the spirit and increase unity of thought in small guerrilla groups. They also put social pressure on the weakest members to better carry out their mission in training and future combat actions.

These group discussions should give special emphasis to creating a favorable opinion of the movement. Through local and national history, make it clear that the enemy regime is repressive and illegal since it does not represent the people. Point out examples of how the government has harmed the people or treated them unjustly. The unification of the nation is the goal. The whole nation will benefit when the resistance wins.

Always include a local focus. International matters are explained only within the context of local events.

Group discussion help each guerrilla learn the need for good behavior to win the support of the population. Discussion guides need to convince the guerrillas that the attitude and opinion of the population play a decisive role, because victory is impossible without popular support.

Running Group Discussions

The guerrilla force is divided into squads for group discussions, including command and support elements, whenever the tactical situation permits it. Small units should be kept together when these groups are designated.

A political cadre is assigned to each group to guide the discussion. The squad leader should help the cadre foster study and the expression of thoughts. If there are not enough political cadres for each squad, leaders should guide the discussions and the available cadres should visit alternate groups.

The cadre or the leader should guide the discussion to cover a number of points and have the group reach a correct conclusion. The guerrillas should feel that it was their own decision. The cadre could also serve as a private teacher.

The political cadre will, at the end of every discussion, make a summary of the principal points, leading to the correct conclusions. Any serious difference with the objectives of the movement should be noted by the cadre and reported to the commander of the force. If necessary, a combined group meeting will be held and a team of political cadres will explain and rectify the misunderstanding.

165

The political cadres should live, eat, and work with the guerrillas. If possible, they should fight at their side. This will foster understanding and the spirit of cooperation that will later help in the discussion and exchange of ideas.

Group discussions should be held in towns and areas of operations whenever possible and not be limited to camps or bases. This is done to emphasize the revolutionary nature of the struggle and to demonstrate that the guerrillas identify with the objectives of the people. The guerrilla projects himself toward the people, as the political cadre does toward the guerrilla.

Dynamics of Group Discussions

Organize discussion groups at the post or squad level. A cadre cannot be sure of the comprehension and acceptance of the concepts and conclusions by members of a large group. With only 10 men to deal with, it is easier to control the situation. In this way, all students will participate in political exchange. Special attention will be given to the individual ability to discuss the objectives of the struggle.

Combine the different points of view and reach an opinion or common conclusion. This is the most difficult task of a political cadre. After the group discusses the democratic objectives of the movement, the chief of the team of political cadres should combine the conclusions of individual groups in a general summary. At a meeting with all the discussion groups, the cadre should provide the principal points, and the guerrillas will have the opportunity to clarify or modify their points of view. To carry this out, the conclusions will be summarized in the form of slogans wherever possible.

Be honest about the national and local problems of the struggle. The political cadres should always be prepared to discuss solutions to the problems observed by the guerrillas. During the discussions, the guerrillas should be guided by the following three principles: freedom of thought, freedom

166

of expression, and concentration of thoughts on the objectives of the struggle.

The desired result is a well-informed guerrilla who can justify all of his acts in a persuasive manner whenever he is in contact with any member of the population, and especially with himself and with his guerrilla companions by facing the challenges of guerrilla warfare.

Camp Procedures

Encamping the guerrilla units gives greater motivation, reduces distractions, and increases the spirit of cooperation between small units. The squad chief will choose the appropriate ground for camping and establish the regular camp procedure. He will give his men such responsibilities as:

- Cleaning the camp area.
- Providing adequate drainage in case of rain.
- Building trenches or holes for marksmen.
- Building a stove by making small trenches and putting rocks in place.
- Building a wind-breaking wall, which will be covered on the sides and on the top with branches and leaves of the same vegetation in the area. This will camouflage it from aerial visibility or enemy patrols.
- Constructing a latrine and a hole where waste and garbage will be buried.
- Positioning a watchman at the places of access at a prudent distance, where the shout of alarm can be heard. At the same time, the password will be established, which should be changed every 24 hours.

The commander should establish ahead of time an alternate meeting point in case the camp has to be abandoned in a hurry. He should establish a third meeting point in case they cannot meet at the established point at a particular time.

167

These and other procedures contribute to the motivation of the guerrilla and improve the spirit of cooperation in the unit. The shared sense of danger, insecurity, anxiety, and other daily concerns result in tangible evidence of belonging to an order, which helps keep up spirit and morale.

Having broken camp with the effort and cooperation of everyone strengthens the spirit of the group. The guerrilla will be inclined toward unity of thought.

Interaction with the People

The enemy government should be identified as the number one enemy of the people and only secondary as a threat against the guerrilla forces. It is not recommended to speak of military tactical plans in discussions with civilians.

Whenever there is a chance, members who have a high political awareness should be sent to civilian areas to persuade the people that the following principles and goals are followed by the guerrillas:

- Maintain respect for human rights and others' property.
- Help the people in community work.
- Protect the people from enemy aggression.
- Teach the people environmental hygiene, how to read, etc.

This attitude will win the sympathy of the people for the movement, and they will immediately become one with the resistance and provide logistical support, intelligence information, and participation in combat.

ARMED PROPAGANDA

Armed propaganda is frequently misunderstood as being a show of force to the people. In reality it does not include compulsion, but the guerrilla should know the principles and methods of this tactic.

Close Identification with the People

Armed propaganda includes all acts carried out by an armed force whose results improve the attitude of the people toward the force. It does *not* include forced indoctrination. It is carried out by a close identification with the people on any occasion. Specifically, the guerrillas should put aside weapons and work side by side with the people: building, fishing, repairing roofs, transporting water, etc. When working with the people, the guerrillas can use slogans such as "many hands doing small things, but doing them together." By doing so, they can establish a strong tie with the people and generate popular support for the movement.

During patrols and other operations around or in the midst of villages, each guerrilla should be respectful and courteous to the people. He should move with care and always be prepared to fight if necessary, but he should not always treat all the people with suspicion or hostility. Even in war, it is possible to smile, laugh, and greet people.

In appropriate places and situations, such as during a rest stop on a march, the guerrillas can explain the operation of weapons to the youths and young men, who are potential recruits for the resistance forces. They can show them an unloaded rifle and how it is used, and let them aim at imaginary targets.

The guerrillas should always be prepared to explain to the people the reason for the weapons. Simple slogans work best, such as:

- "The weapons will be for winning freedom; they are for you."
- "With weapons we can impose demands such as hospitals, schools, better roads, and social services for the people, for you."
- "Our weapons are, in truth, the weapons of the people, yours."

169

- "With weapons we can change the government and return it to the people so that we will all have economic opportunities."

All of this is designed to create an identification of the people with the weapons and the guerrillas who carry them. They should make the people feel that the weapons help them and protect them from an unjust, totalitarian, repressive regime that is indifferent to their needs.

Implicit and Explicit Terror

An armed guerrilla force always involves implicit terror, because the population, without saying it aloud, feels terror that the weapons may be used against them. However, if the terror does not become explicit, positive results can be expected.

In a revolution, the individual lives under a constant threat of physical harm. If the government cannot put an end to guerrilla activities, the population will lose confidence in it, as it has the inherent mission of guaranteeing the safety of citizens. However, the guerrillas should be careful not to resort to explicit terror because this would result in a loss of popular support.

In the words of a leader of the Huk guerrilla movement of the Philippine Islands, "The population is always impressed by weapons, not by the terror that they cause, but rather by a sensation of strength/force. The resistance must appear before the people, giving them the message of the struggle." This is the essence of armed propaganda.

An armed guerrilla force can occupy an entire town or small city that is neutral or relatively passive in the conflict. In order to conduct the armed propaganda in an effective manner, the following should be carried out simultaneously:

- Destroy the military or police installations.

- Cut off all outside lines of communications: cables, radio, messengers.
- Set up ambushes at all possible entry routes in order to delay reinforcements.
- Kidnap all officials or agents of the enemy government and replace them in public places with guerrillas or trustworthy civilians.
- Indicate to the population that, at meetings or in private discussion, they can give the names of the enemy government's informants, who will be removed together with the other officials of the government of repression.
- Establish a public tribunal, and gather the population for this event.
- Shame, ridicule, and humiliate the personal symbols of the government of repression in the presence of the people
- Foster popular participation by having guerrillas within the multitude, shouting slogans and jeers.
- Reduce the influence of individuals associated with the regime by pointing out their weaknesses, then take them out of the town without damaging them publicly.
- Mix the guerrillas within the population and have them display very good conduct.
- Pay with cash for any article.
- Carry out face-to-face persuasion about the struggle.
- Pay courtesy visits to prominent persons such as doctors, priests, and teachers.
- Instruct the population that when the repressive government forces interrogate them at the end of the operation, they may reveal everything. For example, they can identify the type of weapons they use, how many men arrived, from what direction they came and in what direction they left—in short, everything. This is to show that the guerrillas are not afraid of the enemy.
- Conclude meetings with a speech by one of the most dynamic political cadres, which includes explicit refer-

171

ences to the fact that the enemies of the people—the officials or enemy agents—must not be mistreated in spite of their criminal acts, even though the guerrilla force may have suffered casualties, and that this is done due to the generosity of the guerrillas.

- Give a declaration of gratitude for the hospitality of the population. Let them know that the risks that they will run when the enemy returns are greatly appreciated.
- Emphasize the fact that the enemy regime—although it exploits the people with taxes and controls money, food, and all aspects of public life—will not be able to resist the attacks of the guerrilla forces.
- Emphasize that the guerrilla force is fighting for the freedom of our homeland and to establish a people's government.
- Promise the people that you will return to ensure that the repressive enemy regime will not be able to hinder the guerrillas from integrating with the population.

Once again, armed propaganda in populated areas does not give the impression that weapons are the power of the guerrillas over the people but that the weapons are the strength of the people against a repressive regime. Whenever it is necessary to use armed force in an occupation or visit to a town or village, guerrillas should make sure that they:

- Explain to the population that this is being done to protect them, the people, and not the guerrillas.
- Admit frankly and publicly that this is an act of the democratic guerrilla movement, with appropriate explanations.
- State that this action, although it is not desirable, is necessary because the final objective of the insurrection is a free and democratic society, where acts of force are not necessary.

If one of the advanced posts has to fire on a citizen who was trying to leave the town or city in which the guerrillas are carrying out armed propaganda, the following is recommended:

- Explain that if that citizen had managed to escape, he would have alerted the enemy, and they would carry out acts of reprisal such as rape and pillage for having given attention and hospitality to the guerrillas.
- Make the town see that he was an enemy of the people, and that they shot him because they recognize as their first duty the protection of citizens.
- State that the guerrilla who fired the shot tried to detain the informant without firing because he, like all guerrillas, espouses nonviolence. Firing at the enemy informant, although it was done against his own will, was necessary to prevent the repression of the government against innocent people.
- Make the population see that it was the repressive system of the regime that was the cause of this situation and that really killed the informer.
- Convince the population that if the regime had ended the repression, the guerrillas would not have had to brandish arms against their brothers, which goes against their sentiments. This death would have been avoided if justice and freedom existed, which is exactly the objective of the democratic guerrilla.

Selective Use of Violence for Propagandistic Effects

It is possible to neutralize carefully selected targets, such as judges, policemen, state security officials, and tax collectors. For psychological purposes, it is necessary to gather together the people so that they will take part in the act and formulate accusations against the oppressor.

The person should be chosen on the basis of the spontaneous hostility that the majority of the population feels toward him, relative difficulty of controlling the person who

173

will replace the target, degree of violence acceptable to the population, and degree of predictable reprisal by the enemy on the population or other individuals in the area.

Use rejection or hatred of the target by the majority of the population to stir them up and make them see all the negative and hostile actions of the individual against them. If the majority of the people give their support to the target, however, do not try to change these sentiments through provocation.

The mission to neutralize the individual should be followed by an extensive explanation why it was necessary for the good of the people. Furthermore, explain that enemy retaliation is unjust, indiscriminate, and, above all, a justification for the execution of this mission. Carefully monitor the reaction of the people and try to control it, making sure that their reaction is beneficial toward the resistance.

ARMED PROPAGANDA TEAMS

The commanders of a guerrilla force will be able to obtain maximum psychological results from an armed propaganda program. This section will inform the guerrilla student as to what armed propaganda teams are in the environment of guerrilla warfare.

Combination: Political Awareness and Armed Propaganda

Armed propaganda teams combine political awareness building with armed propaganda. It is carried out by carefully selected guerrillas, preferably with experience in combat.

The selection of the staff is more important than the training, because guerrilla cadres cannot be trained to show the sensations of ardor and fervor, which are essential for person-to-person persuasion. It is more important to choose people who are intellectually agile.

An armed propaganda team includes 6 to 10 members. This number is ideal, since it encourages more camaraderie, solidarity, and group spirit. Also, the themes to deal with are assimilated more rapidly and the members react more rapidly to unforeseen situations.

The leader of the group should be the most highly motivated politically and the most effective in face-to-face persuasion. Position or rank are not factors.

Recruitment for guerrilla cadres comes from the same social groups to whom the psychological campaign is directed, such as peasants, students, professionals, housewives, etc. The various cadres will point out to their peers the restrictions and injustices the enemy has placed upon their social group.

Target populations for the armed propaganda teams are chosen because they are part of the operational area and not for their size or amount of land. The objective should be the people and not the territorial area.

The target groups for the armed propaganda teams are not the persons with sophisticated political knowledge but rather those whose opinions are formed from what they see and hear.

The training of guerrillas for armed propaganda teams emphasizes the method and not the content. An intensive two-week training period is sufficient if the recruitment is done in the form indicated. If the wrong candidate is recruited, no matter how good the training, he will not yield a very good result.

Team discussions are used during training, alternating the person who leads the discussion. A different theme is presented each day. The themes should refer to the conditions of the area and the meaning that they have for the inhabitants, such as crops, fertilizers, seeds, irrigation, etc.

Persuasion Principles

It is a principle of psychology that humans have the tendency to form personal associations from "we" and "the others," or "we" and "they," "friends" and "enemies," "fellow countrymen" and "foreigners." The armed propaganda team can use this principle in its activities, so that it is obvious that the "exterior" groups ("false" groups) are those of the enemy regime, and that the "interior" groups ("true" groups) are the freedom fighters. The team should introduce this to the people in a subtle manner so that these feelings seem to be born spontaneously.

It is a principle of political science that it is easier to persuade people to vote *against* something or someone than to persuade them to vote in favor of something or someone. The armed propaganda teams should ensure that their campaign is directed specifically against the government or its sympathizers, since the people need specific targets for their frustrations.

Another principle of sociology is that humans forge or change opinions from two sources—primarily through association with family, comrades, and intimate friends, and secondarily through distant associations such as acquaintances in churches, clubs, committees, labor unions, and governmental organizations. The armed propaganda teams should join the first groups in order to persuade the people to follow the policies of the resistance movement, because it is from these groups where opinions or changes of opinion come.

Psychological tactics should retain a degree of flexibility within a general plan, permitting a continuous and immediate adjustment of the message and ensuring that it has maximum impact on the target group at the moment in which the group is most susceptible.

Techniques of Persuasion in Talks and Speeches

Be simple and concise. Avoid the use of difficult words or expressions, and use popular words and expressions, i.e., the language of the people. It is important to use oratory to make

176

people understand the reason for the struggle rather than to show off knowledge.

Use lively and realistic examples. Avoid abstract concepts. Instead, give concrete examples such as children playing, horses galloping, birds in flight, etc.

Use gestures to communicate such as expressive hands, back movements, facial expressions, focusing of a look, and other aspects of body language.

Use the appropriate tone of voice. If, when addressing the people, you talk about happiness, a happy tone should be used. If you talk of something sad, the tone of voice should be sad. When talking of a heroic or brave act, the voice should be animated.

Above all, be natural. Imitation of others should be avoided, since people can distinguish a fake easily. The individual personality should be projected when addressing people.

"Eyes and Ears" Within the Population

The amount of information that will be generated by the armed propaganda teams will allow the resistance to cover a large area for intelligence. The teams will become the eyes and ears of the movement within the population. However, it is necessary to emphasize that the first mission of the armed propaganda team is to carry out psychological operations, not to obtain data for intelligence.

The armed propaganda teams are able to do what others in a guerrilla campaign cannot do: determine personally the development or deterioration of popular support and the sympathy or hostility that the people feel toward the movement. They report to their superior the reaction of the people to radio broadcasts, provocative flyers, or any other means of propaganda of the resistance. With the intelligence reports supplied by the armed propaganda teams, the commanders will be able to have exact knowledge of the popular support, which they will make use of in their operations.

177

Team Security

Each armed propaganda team will be able to cover several towns. The team should always move in a covert manner within the towns of their area. They should vary their route radically, but not their itinerary. This is so that the inhabitants who are cooperating will be dependent on their itinerary, i.e., the hour in which they can make contact with them to give them information. The method of making contact should be covert. There should be predetermined all-clear signals. These signals should be placed at a specific, predetermined time, although the time for the signals should be varied.

When the unit makes contact with the designated representative of the town, one person goes forward, covered by a sniper, until the all-clear is given.

Prior to making contact, the team should keep the community under surveillance to see if there is any unusual activity or if there has been any changes that warrant concern.

No more than three consecutive days should be spent in a town. This limit has obvious tactical advantages, but it also has a psychological effect on the people who see the team as a source of current and up-to-date information. Also, more than three days can overexpose the armed propaganda team and cause a negative reaction.

Basic tactical precautions should be taken. This is necessary for the team's security. When it is carried out discreetly, it increases the respect of the people and increases the team's credibility. The danger of betrayal or an ambush can be neutralized, for instance, by varying the itinerary a little, using different routes, and arriving or leaving without warning.

Mixing with the Target Groups

Although meetings may be held with the general population, the armed propaganda teams should recognize and keep in contact with the target groups, mixing with them before, during, and after the meeting.

178

In the first visits, the guerrilla cadres are courteous and humble. They contribute to the improvement of the life of the inhabitants—work in the fields, help to repair fences, help with vaccination of their animals, teach them to read.

In his free time, the guerrilla should mix in with the groups and participate with them in pastoral activities, parties, birthdays, and even wakes or burials. He tries to converse with both adults and adolescents. He tries to penetrate to the heart of the family in order to win the acceptance and trust of all of the residents of that sector.

The cadres should not make mention of their political ideology during the first phase of identification with the people. They should orient their talks to things that are pleasing to the people, trying to be as simple as possible in order to be understood.

The cadres should gradually mix ideological training in with things like folk songs. At the same time he can tell stories that have some attraction, such as the heroic acts of their ancestors. He should also tell stories of heroism of the combatants in the present struggle so that listeners try to imitate them. It is important to let them know that there are other countries in the world where freedom and democracy cause those governing to be concerned with the well-being of the people so that children have medical care and free education; everyone has work and food; all freedoms such as religion, association, and expression are respected; and where the greatest objective of the government is to keep its people happy and prosperous.

The tactical objectives for identification with the people are to establish tight relations with the people through their customs, determine the basic needs and desires of the different target groups, discover the weaknesses of governmental control, and sow the seed of democratic revolution.

For economic interest groups such as small businessmen and farmers, it should be emphasized that their potential

179

progress is limited by the enemy government, that resources are scarcer and scarcer, that profits are minimal, taxes high, and so on.

For the elements ambitious for power and social positions, it should be emphasized that they will never be able to belong to the governmental social class, since they are hermetic in their circle of command. For example, the enemy leaders do not allow other persons to participate in the government, and they hinder economic and social development of those unlike them, which is unjust and arbitrary.

Social and intellectual criticisms should be directed at professors, teachers, priests, missionaries, students, and others. Make them see that their writings, commentaries, and conversations are censored, which does not make it possible to correct the problems addressed.

Once the needs and frustrations of the target groups have been determined, the hostility of the people will become more direct against the current regime and its system of repression. The people will be made to see that once this system is eliminated, the cause of their frustrations will be eliminated. It should be shown that supporting the insurrection is really supporting their own desires, since the movement is aimed at the elimination of these specific problems.

Avoiding Conflict

As a general rule, the armed propaganda teams should avoid participating in combat. However, if this is not possible, they should react as a guerrilla unit with hit and run tactics, causing the enemy the greatest amount of casualties with aggressive assault fire, recovering enemy weapons, and withdrawing rapidly.

One exception to the rule to avoid combat is when they are challenged by hostile actions while in the town, whether by an individual or by an enemy team. The hostility of one or two men can be overcome by eliminating them in a rapid and effective manner. (This is the most common danger.) When the enemy is equal in number, there should be an immediate retreat, and then the enemy should be ambushed or eliminated by sniper fire.

In any of these cases, the armed propaganda team should not turn the town into a battleground. Generally, the guerrillas will obtain greater respect from the population if they carry out appropriate maneuvers instead of endangering lives or destroying houses in an encounter with the enemy within the town.

A Comprehensive Team Program

Psychological operations through armed propaganda teams include the infiltration of key guerrilla communicators into the population instead of sending messages to them through outside sources, thus creating a "mobile infrastructure."

A mobile infrastructure is a cadre of the armed propaganda team that keeps in touch with six or more populations. Their goal is to integrate the populations in the guerrilla movement at the appropriate time. In this way, an armed propaganda team program builds a constant source of data gathering throughout the operational area. It is also a means for developing and increasing popular support for recruiting new members and obtaining provisions.

In addition, an armed propaganda team program helps to expand the guerrilla movement by penetrating areas that are not under the control of the combat units. In this way, through an exact evaluation of the enemy combat units in the area, the guerrilla commanders will be able to plan operations more precisely since they will have certain knowledge of existing conditions.

This type of operation is similar to the "Fifth Column" that was used in the first part of World War II. Through infiltration and subversion tactics, the Germans were able to penetrate the target countries before invading. They man-

181

aged to capture Poland, Belgium, Holland, and France in a month, and Norway in a week. The effectiveness of this tactic has been clearly demonstrated in several wars and can be used effectively by the resistance.

In the same way that the security elements are the eyes and ears of a patrol or a column on the march, the armed propaganda teams are the "antennas" of the movement because they find and exploit the sociopolitical weaknesses in the target society, making a successful operation possible. The activities of the armed propaganda teams run some risks, but no more than any other guerrilla activity. However, they are essential for the success of the struggle.

DEVELOPMENT AND CONTROL OF FRONT ORGANIZATIONS

The development and control of front organizations (or "facade" organizations) is an essential process in the guerrilla effort. It is generally an aspect of urban guerrilla warfare, but it should advance parallel to the campaign in the rural area. This section will give the guerrilla student an understanding of the development and control of front organizations in guerrilla warfare.

Initial Recruitment

The initial recruitment to the movement, if it is involuntary, is carried out through several "private" consultations with a cadre (i.e., without his knowing that he is talking to a recruiter). If the recruit is agreeable, he or she is trained to carry out simple missions.

When the guerrillas begin to carry out missions of armed propaganda and conduct regular visits to the towns, the cadres will provide the guerrillas with the names of persons who can be recruited. The recruitment, which will be voluntary, is done through guerrilla leaders or political cadres.

After a group of voluntary recruits has been developed and their trustworthiness established by their carrying out small missions successfully, they will be taught on how to increase the group by recruiting in specific target groups in accordance with the following procedure.

From among their acquaintances or through observation of target groups—political parties, workers' unions, youth groups, agrarian associations, etc.—they are told to find out the personal habits, preferences, biases, and weaknesses of recruitable individuals.

Next, they make an approach through an acquaintance and, if possible, develop a friendship. They begin to work on him through his preferences or weaknesses. It might entail inviting him for lunch or dinner in the restaurant of his choice, or having a drink in his favorite bar.

If, in informal conversation, the target seems susceptible to voluntary recruitment based on his beliefs and personal values, the political cadre assigned to carry out the recruitment is notified. The original contact will tell the cadre all he knows of the prospective recruit and the style of persuasion to be used. He will then introduce the two.

If the target does not seem to be susceptible to voluntary recruitment, seemingly casual meetings can be arranged with the guerrilla leaders or political cadres (unknown by the target until that moment). The meetings will be held so that "other persons" know that the target is attending them, whether they see him arrive at a particular house, seated at the table in a particular bar, or even seated on a park bench. The target, then, is faced with the appearance of his participation in the insurrection, and he will be told that if he fails to cooperate or carry out future orders, he will be subjected to reprisals by the police or soldiers of the regime. Explain to the target that the resistance knows where he lives and works as well as where his friends and family live and work.

When it becomes necessary, the police can be notified

through a letter containing false statements by citizens who are not involved in the movement. This letter will denounce the uncooperative target. Care should be taken that the person who attempted to recruit him is not discovered.

The involvement and handing of more recruits is done gradually on a wider and wider scale. This should be a gradual process in order to prevent compromise by inexperienced individuals who have been assigned difficult or dangerous missions too early.

Using this recruitment technique, the guerrillas should be able to successfully infiltrate any key target group in the regime in order to improve internal control of the enemy structure.

Established Citizens, Subjective Internal Control

Established citizens such as doctors, lawyers, businessmen, landholders, and minor state officials will be recruited to the movement and used for subjective internal control of groups and associations to which they belong. Once the recruitment has progressed to a certain level, specific instructions can be given to internal cadres to begin to influence their groups.

The process is simple and requires identification of some theme, word, or thought related to the objective of persuasion. The cadre then must emphasize this theme, word, or thought in the discussions or meetings of the target group through casual commentary.

For example, economic interest groups are motivated by profit and generally feel that the system hinders the use of their abilities in some way, such as by taxes, import-export tariffs, or transportation costs. The cadre in charge will increase this feeling of frustration.

Political aspirants feel that the system discriminates against them unfairly because the enemy regime does not allow elections. The cadres should focus political discussions towards this frustration.

184

Intellectual social critics such as professors, teachers, priests, and missionaries generally feel that the government ignores their valid criticism or censors their comments unjustly, especially in a revolutionary environment. This can easily be shown to be an inherent injustice of the system.

For all the target groups, after they have established their frustrations, the hostility toward the obstacles to their aspirations is gradually transferred to the current regime and its system of repression.

The guerrilla cadre moving among the target groups should always maintain a low profile so that the development of hostile feelings toward the enemy regime seems to come spontaneously from the members of the group and not from the suggestions of the cadres. This is internal subjective control.

Antigovernment hostility should be generalized and not necessarily in favor of the resistance. If a group does develop sympathy in their favor, it can be utilized, but the main objective is to precondition the target groups for fusion into mass organizations later in the operation.

Organizations of Cells for Security

The three-person cell is the basic element of the movement. It holds frequent meetings to receive orders and pass information to the cell leader. These meetings are also very important for mutually reinforcing the morale of the cell members.

The coordination of the three-member cell provides a security net for two-way communication, each member having contact with only one operational cell. The members will not reveal at the cell coordination meetings the identity of their contact in an operational cell; they will reveal only the nature of the activity in which the cell is involved, e.g., political party work, medical association work, etc.

There is no hierarchy in cells outside of an element of coordination, e.g., who is the leader, who will have direct covert contact with the guerrilla commander in the operational area. For every three operational cells, a coordination cell is needed.

Fusion in a Cover Organization

The fusion of associations and other groups through internal subjective control occurs in the final stages of the operation. It is tightly connected with mass meetings.

When the insurrection has expanded sufficiently, armed propaganda missions are carried out on a large scale. Propaganda teams will have clearly developed open support from well-infiltrated and preconditioned target groups. At the point at which mass meetings are held, the internal cadres should begin discussions for the fusion of forces into an organization that will be a cover group for the movement.

Target groups will be aware that other groups are developing greater hostility toward the government, police, and traditional bases of authority. Guerrilla cadres in particular groups—teachers, for example—will cultivate this awareness-building, making comments such as, "So-and-so, who is a farmer, said that the members of his cooperative believe that the new economic policy is absurd, poorly planned, and unfair to them."

When the awareness-building is increased to the point where other groups feel hostility toward the regime, group discussions are held openly and the movement will receive reports that the majority of their operatives are united in common, greater hostility against the regime. Then the order to fuse will come.

The fusion into a cover front is carried out as follows. Internal cadres meet with leaders and others at organized meetings chaired by the chief of the movement. Two or three escorts can assist the guerrilla cadre if it becomes necessary.

A joint communiqué on this meeting is published, announcing the creation of the cover front, including names and signatures of the participants and names of the organizations that they represent.

After releasing this communiqué, mass meetings are initiated that have as a goal the destruction of enemy control.

The development and control of cover organizations in a guerrilla war will give the movement the ability to create a whiplash effect within the population when the order for fusion is given. When infiltration and internal subjective control have been developed parallel with other guerrilla activities, a guerrilla commander will be able to shake up the enemy structure and replace it.

CONTROL OF MASS
CONCENTRATIONS AND MEETINGS

In the later stages of a guerrilla war, mass concentrations and meetings are a powerful psychological tool for carrying out the mission. This section will give the guerrilla student training on techniques for controlling mass concentrations and meetings in guerrilla warfare.

Infiltration of Guerrilla Cadres

As discussed, guerrilla cadres are infiltrated into workers' unions, student groups, peasant organizations, etc., preconditioning these groups for later violent behavior. The psychological war team should foster a hostile environment among the target groups so that, at the decisive moment, they can turn that furor into violence.

The basic objective of a preconditioning campaign is to create a negative image of the common enemy, e.g., the government is corrupt, the police mistreat the people. Make it plain to the people that they have become slaves and that they are being exploited by privileged military and political groups.

These preconditioning campaigns must be aimed at the

187

politicians, professionals, students, laborers, unemployed, ethnic minorities, and any other segment of society that is vulnerable or recruitable. This also includes the general masses and sympathizers of the movement.

The cadres eventually will create temporary compulsive obsessions in places of public concentrations, constantly hammering away at the desired themes developed at earlier group gatherings, in informal conversations, and through brochures, flyers, and editorial articles both on the radio and in newspapers.

Selection of Appropriate Slogans

The leaders of the movement develop slogans in accordance with the circumstances. The aim is to mobilize the masses at the highest emotional level.

When the mass uprising is being developed, slogans should make partial demands, such as, "We want food," "We want freedom of worship," or "We want union freedom." If the enemy is lacking in organization and command, and the people are sufficiently riled up, the agitators can take advantage of the situation and raise the tone of the rallying slogans, taking them to the most strident point.

If the masses are not emotionally aroused, the agitators will continue with the partial slogans, and the demands will remain based on daily needs, but always chaining them to the goals of the movement.

An example of the need to use simple slogans is that few people think in terms of millions of dollars, but any citizen, however humble, understands that a pair of shoes is necessary. The goals of the movement are of an ideological nature, but the agitators must realize that food and daily necessities pull the people along.

Creation of Nuclei

This involves the mobilization of a specific number of

agitators who will inevitably attract an equal number of curious persons who seek adventure as well as those unhappy with the government. Each guerrilla subunit will be assigned specific tasks and missions that they should carry out.

The cadres will be mobilized in the largest number possible together with people who have been affected by the government, whether their possessions have been stolen from them, they have been incarcerated or tortured, or suffered from any other type of aggression. They will be directed toward the areas where the hostile elements of the government are based. An effort should be made to arm them with placards, clubs, iron rods, and small firearms if possible, which they will keep hidden.

Designated cadres will arrange ahead of time transportation for the participants in private or public vehicles, boats, or any other type of transportation. Other cadres will be designated to design placards, flags, and banners with different slogans or key words, whether they be partial or of the most radical type. Still other cadres will prepare flyers, posters, signs, and pamphlets. This material will contain instructions for the participants as well as slogans against the regime.

If the government takes violent action against the demonstrators, this action is reported to the people and condemned. If any persons are killed, they are made into martyrs and further demonstrations are held in their memory.

Leading an Uprising at Mass Meetings

A small group of guerrillas infiltrated within the masses will agitate them, giving the impression that there are many of them and that they have popular backing. Using these tactics, 200-300 agitators can create a demonstration in which 10,000-20,000 persons take part.

189

Outside Command

This element stays out of all activity. They are located so that they can observe and control the development of the planned events. This could be a church tower, tall building, large tree, highest level of a stadium or auditorium, or any other high place.

Inside Command

This element remains within the multitude. The leaders of this element *must* be protected. Placards or large signs should be used to designate their command post and provide signals to the subunits. In this way the commander will be able to send orders to change passwords or slogans, handle any unforeseen thing, and even incite violence if he desires it. This element will avoid placing itself in places where fights or other incidents are taking place.

At this stage, key cadres place themselves in visible places such as by signs, lampposts, and other places that stand out. They should also avoid disturbances once they have started them.

Defense Posts

These elements act as bodyguards, protecting the chief from the police and the army or helping him to escape if it should be necessary. They should be highly disciplined and react only upon a verbal order from the chief.

If the chief participates in a religious ceremony, funeral, or any other organized activity, the bodyguards will remain very close to him or to his placard/banner carriers in order to give them full protection. Bodyguards should be guerrilla combatants in civilian clothes or hired recruits who are sympathizers to the struggle.

Messengers

Messengers should remain near the leaders, transmitting

orders between the inside and outside commands. They use radios, telephones, bicycles, motorcycles, or cars, or move on foot or horseback, taking paths or trails to shorten distances. Adolescents (male and female) are ideal for this mission.

Shock Troops

These men should be equipped with weapons (knives, razors, chains, clubs, bludgeons) and should march slightly behind the forward participants. They should keep their weapons hidden and enter into action only as reinforcements if the guerrilla agitators are attacked by the police. They will enter the scene quickly, violently, and by surprise in order to distract the authorities so the inside command can withdraw or escape.

Banner and Placard Carriers

Banners and placards used in demonstrations will express the protests of the people via slogans and key words. When the concentration reaches its highest level of emotion and discontent, the infiltrated persons will pull out the placards, which have remained hidden until this time. The one responsible for this mission will instruct the agitators ahead of time to keep near the placard of any contrary element. In that way, the commander will know where the agitators are and will be able to send orders to change slogans and eventually incite violence if he wishes.

Agitators of Rallying Cries and Applause

These people are trained to use carefully chosen rallying cries such as, "We are hungry," "We want bread," and "We don't want _____ ." Their technique for agitating the masses is quite similar to cheerleaders at high school football games. The objective is to become more adept at stirring things up and not just to shout rallying cries.

MASSIVE SUPPORT THROUGH PSYCHOLOGICAL OPERATIONS

The following paragraphs summarize the key points of a successful psychological operation.

Motivation as Combatant-Propagandist. Every member of the struggle should know that his political mission is as important as, if not more important than, his tactical mission.

Armed Propaganda. Armed propaganda in rural villages, small towns, and cities should give the impression that the weapons of the resistance are not for exercising power over the people but rather for protecting the people against the government of oppression.

Armed Propaganda Teams. Armed propaganda teams will combine political awareness building with propaganda within the population.

Cover Organizations. Through internal subjective control, the fusion of several organizations and associations recognized by the government occurs in the final stages of the operation in close coordination with mass meetings.

Control of Mass Demonstrations. The appearance of a spontaneous demonstration must be maintained, but the participants will be controlled by the agitators.

Too often we see guerrilla warfare only from the point of view of combat actions. This view is erroneous and extremely dangerous. Combat actions are not the key to victory in guerrilla warfare but rather are only one part of the effort. There is no priority in any of the efforts; they should progress in a parallel manner. The emphasis or exclusion of any of these efforts could bring about serious difficulties and even failure. The history of revolutionary wars has shown this to be true.

OTHER PSYCHOLOGICAL OPERATIONS

Enemy Military Forces. Enemy military forces may be of the same nationality as the guerrillas. If so, radio broadcasts telling them news of the recent successes of the resistance, why the people are fighting the government, and how the government is lying to them helps to make them feel isolated and plays on their insecurities. Determining the sources of frustration for the enemy troops can help dictate what amplifications and focus is needed.

Through radio broadcasts, leaflets, loud speakers, etc., the enemy soldier is promised safety if he deserts. This promise must be kept. When an enemy soldier turns himself over to the resistance, he is requested to make it known to his former comrades that he has been treated well.

Raids, ambushes, sniper operations, random mortar attacks, etc. have a demoralizing effect on enemy soldiers.

Civilian Population. One of the ways the population receives news and information from the resistance is through radio. Broadcasts by respected former leaders denouncing the government and telling the people what the enemy government is doing to them is often a good approach. Another method is through leaflets or newspapers.

APPENDIX

The purpose of this appendix is to complement the guidelines and recommendations under "Techniques of Persuasion in Talks and Speeches" to improve the ability to organize and express thoughts. After all, oratory is one of the most valuable resources for exercising leadership and can be an extraordinary political tool.

The Audience
The orator and his audience share the same time and space. Therefore, every speech should be a different experience at "that" moment which the audience is experiencing. So the audience must be considered as "a state of mind."

193

Happiness, sadness, anger, fear, etc., are states of mind that must be considered to exist in the audience.

Human beings act in accordance with their thoughts and sentiments and respond to the stimuli of ideas and emotions. In that way there exists only two possible focuses for any speech: the concrete, based on rational appeals (thought), and the idealized, presented with emotional appeals (sentiment).

The orator must be sensitive to the existing mass sentiment, but he must also keep his judgment to be able to lead and control the feelings of an audience.

Qualities of a Speech

Political oratory usually fulfills one of three objectives: to instruct, persuade, or move. It does so by urging, ordering, questioning, and responding.

There are specific qualities of an effective speech, especially a political speech in the context of a psychological action.

First, be brief and concise. Five minutes is about right. Stay centered on the theme. The speech should be structured by a set of organized ideas that converge on the theme. If there is no guiding ideas or themes in a speech, it could result in confusion and dispersion of the audience.

The ideas presented should be logical and easily acceptable. Never challenge logic in the mind of the audience, since credibility is lost immediately. As far as possible, it is recommended that all speeches be based on symbolism. For example, "Those government officials get rich and are thieves."

Structure of a Speech

Absolute improvisation does not exist in oratory. All orators have a mental plan that allows them to organize and present their ideas and concepts rapidly. With practice it is possible to do this almost simultaneously with the expression of the words.

194

For those who wish to improve their oratorical abilities, the elements that make up a speech are given below.

Introduction or Preamble

A personal introduction is made, one of the organizations to which he belongs is identified, and/or the reason for his presence is given. In the introduction it is important to make an impact and provoke interest among the audience members. You can begin with a famous phrase or a previously prepared slogan, or you can tell a dramatic or humorous story.

Purpose or Enunciation

The subject to be dealt with is defined and explained as a whole or in parts.

Appraisal or Argumentation

Arguments are presented *exactly in this order*. First, the negative arguments against the thesis that is going to be presented, and then the positive arguments that are favorable to the thesis, immediately adding proof or facts that sustain such arguments.

Recapitulation or Conclusion

A short summary is made and the conclusions of the speech are spelled out.

Exhortation

Action by the public is called for, i.e., they are asked in an almost energetic manner to do or not do something.

Some Literary Resources

Although there are specific devices for effective oratory, in truth, oratory has taken a large number of devices from other literary genres, several of which we often use in our daily expressions.

Below are listed some of the literary devices that are used frequently in oratory. Use them moderately, since an orator who overuses literary devices loses impact.

- Repetition of a word at the beginning of each sentence: "Freedom for the poor, freedom for the rich, freedom for all."
- Repetition at the end of every phrase: "Communism tries to dominate everyone, command everyone, and as an absolute tyranny, do away with everyone."
- Repetition of a complete sentence or slogan insistently through the speech: "With God and patriotism we will win this struggle." "With God and patriotism we will overcome Communism."
- Beginning a phrase with the same word that ended the previous one: "We struggle for democracy, democracy and social justice." You can also make a "chain" of such duplications: "Communism transmits the deception of the child to the young man, of the young man to the adult, and of the adult to the old man."
- Word play, where the same words are used with a different meaning: "The greatest wealth of every human being is his own freedom, because slaves will always be poor but we poor can have the wealth of our freedom."
- Similar cadences, through the use of verbs of the same tense and person, or nouns of the same number and case: "Those of us who are struggling, we will be marching because he who perseveres achieves, and he who gives up remains."
- Synonyms, repetition of words with a similar meaning: "We demand a government for all, without exceptions, without omissions."
- Comparison or simile, which sets the relationship of similarity between two or more things: "Because we love Christ, we love his bishops and pastors," and "Free as a bird."

196

- Antithesis, or the use of words, ideas, or phrases of an opposite meaning: "They promised freedom and gave slavery; that they would distribute the wealth, and they have distributed poverty; that they would bring peace, and they have brought war."
- Concession, which is a skillful way to concede something to the adversary in order to better emphasize inappropriate aspects. This is done through the use of such expressions as "but," "however," "although," "nevertheless," and "in spite of the fact that." For example, "The mayor here has been honest, but he is not the one controlling all the money of the nation." It is an effective form of rebuttal when the opinion of the audience is not entirely yours.
- Permission, in which one apparently accedes to something, when in reality it is rejected: "Do not protest, but sabotage them." "Talk quietly, but tell it to everyone."
- Addressing anticipated refutations: "Some will think that these are only promises; they will say, others said the same thing. But no. We are different. We are Christians. We consider God a witness to our words."
- Pretending discretion when something is said with total clarity and indiscretion: "If I were not obligated to keep military secrets, I would tell all of you of the large amount of armaments that we have so that you would feel even more confidence that our victory is assured."
- Asking and giving the answer to the same question: "If they show disrespect for the ministers of God, will they respect us, simple citizens? Never."
- Asking rhetorical questions to show perplexity or an inability to say something, but only as an oratorical recourse: "I am only a peasant. I know little and I will not be able to explain to you the complicated things of politics. Therefore, I talk to you with my heart, with my simple peasant's heart, as we all are."
- Phrases that mean a lot by saying little: "The government

officials have stolen little, just the whole country."
- Irony, getting across exactly the opposite of what one is saying: "The divine mobs that threaten and kill, they are indeed Christians."
- Amplification, presenting an idea from several angles: "Political votes are the power of the people in a democracy. And economic votes are their power in the economy. Buying or not buying something, the majorities decide what should be produced. For something to be produced or to disappear. That is part of economic democracy."
- Entreaty to obtain something: "Lord, free us from the yoke. Give us freedom."
- Expressing a threat: "Let there be a homeland for all or let there be a homeland for no one."
- Presenting a bad wish on the enemy: "Let them drown in the abyss of their own corruption."
- Addressing something supernatural or inanimate as if it were a living being: "Mountains of our homeland, make the seed of freedom grow."
- Asking a question of oneself to give greater emphasis to what is expressed: "If they have already injured the members of my family, my friends, my peasant brothers, do I have any path other than brandishing a weapon?"
- Intentionally leaving a thought incomplete so that mentally the audience completes it: "They promised political pluralism and gave totalitarianism. They promised social justice, and they have increased poverty. They offered freedom of thought, and they have given censorship. Now, what they promise are free elections."

COMBAT IN URBAN AREAS

Guerrilla forces may be required to fight the enemy in urban areas as well as in forests and jungles. As the military situation is developed by the guerrillas, they will take the initiative and start hitting the enemy in villages, towns, and cities.

This chapter only touches upon the basics of urban fighting by a small unit. Much more detailed information can be found in U.S. Army FM 90-10-1 dealing with combat in built-up areas.

HOW TO ATTACK AND CLEAR A BUILDING

Attack one building at a time. The attack leader designates a part of his force as the assault force and the rest as the support force. As a rule, any machine guns and rocket launchers will be assigned to the support force.

A building is attacked in three steps:

- The support force and any indirect fire support isolate the building.

- The assault force enters the building to seize a foothold.
- The assault force clears the building room by room.

To isolate the building, the support force takes up an overwatch position. It fires and adjusts indirect fire to suppress enemy troops in the building and those in nearby buildings who can shoot at the assault force.

The assault forces moves to the building on a covered and concealed route using smoke grenades or smoke pots for additional concealment. The assault force enters the building at the highest level it can, because the ground floor is usually where the enemy places his strongest defenses. The roof is usually weaker than the walls, and it is easier to fight while moving down stairs than up stairs.

If there is no covered route to the roof, the assault force may enter at a lower story or at ground level. In this case, it should seize a foothold quickly, fight to the highest story, and then clear from the top down.

To enter the building, one of the two- or three-man assault parties moves (covered by fire) to the nearest covered area near the point of entry (this may be the outer wall of the building if made of stone). One man throws a grenade into the room. After the explosion, the party enters one at a time overwatched by the rest. The first man enters firing two and three round bursts. He takes a position to cover the whole room. The other men enter the room and make a quick but thorough search. The same procedure is repeated from room to room, floor to floor, until the building is cleared of all enemy. It is then secured. If feasible, it becomes the position from which the assault on the next building is overwatched.

If the building is made of material that can be penetrated by small arms, beware of possible inner fortifications such as sandbags.

The enemy will probably place snipers and rocket

200

launchers on the upper floors. Machine guns will most likely be placed on the ground floor to achieve grazing fire.

Individual positions may be located in the shadows of rooms but use windows as firing ports.

Holes in outer walls close to the ground may be used and concealed by plants.

The enemy may have "mouse holes" cut between rooms and floors to give him covered routes to alternate positions.

Beware of mines outside the building and boobytraps at entrances and windows.

Doors and windows will be covered by fire.

Remember:

- If there is no cover, do not try to climb the side of a building.
- Keep away from the middle of streets.
- Do not silhouette yourself climbing over walls or through windows.
- "Cook-off" a grenade for 2 seconds before throwing it (if U.S. grenades).
- Do not throw grenades up stairs.
- Setting a building on fire may be the best way to neutralize it.
- Buildings may be interlinked by tunnels or sewers.